I0488646

Precipitation and Runoff Simulations of Select Perennial and Ephemeral Watersheds in the Middle Carson River Basin, Eagle, Dayton, and Churchill Valleys, West-Central Nevada

By Anne E. Jeton and Douglas K. Maurer

Prepared in cooperation with the Bureau of Reclamation

Scientific Investigations Report 2011–5066

U.S. Department of the Interior
U.S. Geological Survey

U.S. Department of the Interior
KEN SALAZAR, Secretary

U.S. Geological Survey
Marcia K. McNutt, Director

U.S. Geological Survey, Reston, Virginia: 2011

For more information on the USGS—the Federal source for science about the Earth, its natural and living resources, natural hazards, and the environment, visit http://www.usgs.gov or call 1–888–ASK–USGS.

For an overview of USGS information products, including maps, imagery, and publications, visit http://www.usgs.gov/pubprod

To order this and other USGS information products, visit http://store.usgs.gov

Suggested citation:
Jeton A.E., and Maurer, D.K., 2011, Precipitation and runoff simulations of select perennial and ephemeral watersheds in the middle Carson River basin, Eagle, Dayton, and Churchill Valleys, west-central Nevada: U.S. Geological Survey Scientific Investigations Report 2011–5066, 44 p.

Contents

Figures

Figures—Continued

Tables

Conversion Factors, Datums, and Abbreviations and Acronyms

Conversion Factors

Inch/Pound to SI

Multiply	By	To obtain
Length		
inch (in.)	2.54	centimeter (cm)
inch (in.)	25.4	millimeter (mm)
foot (ft)	0.3048	meter (m)
mile (mi)	1.609	kilometer (km)
Area		
acre	4,047	square meter (m^2)
acre	0.4047	hectare (ha)
acre	0.4047	square hectometer (hm^2)
acre	0.004047	square kilometer (km^2)
square mile (mi^2)	259.0	hectare (ha)
square mile (mi^2)	2.590	square kilometer (km^2)
Volume		
acre-foot (acre-ft)	1,233	cubic meter (m^3)
acre-foot (acre-ft)	0.001233	cubic hectometer (hm^3)
Flow rate		
cubic foot per second (ft^3/s)	0.02832	cubic meter per second (m^3/s)

Temperature in degrees Celsius (°C) may be converted to degrees Fahrenheit (°F) as follows:

$$°F=(1.8×°C)+32$$

Datums

Vertical coordinate information is referenced to the North American Vertical Datum of 1988 (NAVD 88).

Horizontal coordinate information is referenced to the North American Datum of 1983 (NAD 83).

Altitude, as used in this report, refers to distance above the vertical datum.

Conversion Factors, Datums, and Abbreviations and Acronyms—Continued

Abbreviations and Acronyms

DEM	digital elevation model
ET	evapotranspiration
GIS	geographic information system
HA	hydrographic area
HRU	hydrologic response unit
NLCD	national land cover data
PET	potential evapotranspiration
PZM	precipitation-zone method
PRMS	Precipitation-Runoff Modeling System
RAWS	remote automated weather station
RMSE	root mean square error
SNOTEL	snowpack telemetry
STATSGO	State soil geographic database, U.S. Department of Agriculture, National Soil Survey Center
USGS	U.S. Geological Survey

Precipitation and Runoff Simulations of Selected Perennial and Ephemeral Watersheds in the Middle Carson River Basin, Eagle, Dayton, and Churchill Valleys, West-Central Nevada

By Anne E. Jeton and Douglas K. Maurer

Abstract

The effect that land use may have on streamflow in the Carson River, and ultimately its impact on downstream users can be evaluated by simulating precipitation-runoff processes and estimating groundwater inflow in the middle Carson River in west-central Nevada. To address these concerns, the U.S. Geological Survey, in cooperation with the Bureau of Reclamation, began a study in 2008 to evaluate groundwater flow in the Carson River basin extending from Eagle Valley to Churchill Valley, called the middle Carson River basin in this report. This report documents the development and calibration of 12 watershed models and presents model results and the estimated mean annual water budgets for the modeled watersheds. This part of the larger middle Carson River study will provide estimates of runoff tributary to the Carson River and the potential for groundwater inflow (defined here as that component of recharge derived from percolation of excess water from the soil zone to the groundwater reservoir).

The model used for the study was the U.S. Geological Survey's Precipitation-Runoff Modeling System, a physically based, distributed-parameter model designed to simulate precipitation and snowmelt runoff as well as snowpack accumulation and snowmelt processes. Models were developed for 2 perennial watersheds in Eagle Valley having gaged daily mean runoff, Ash Canyon Creek and Clear Creek, and for 10 ephemeral watersheds in the Dayton Valley and Churchill Valley hydrologic areas. Model calibration was constrained by daily mean runoff for the 2 perennial watersheds and for the 10 ephemeral watersheds by limited indirect runoff estimates and by mean annual runoff estimates derived from empirical methods. The models were further constrained by limited climate data adjusted for altitude differences using annual precipitation volumes estimated in a previous study. The calibration periods were water years 1980–2007 for Ash Canyon Creek, and water years 1991–2007 for Clear Creek. To allow for water budget comparisons to the ephemeral models, the two perennial models were then run from 1980 to 2007, the time period constrained somewhat by the later record for the high-altitude climate station used in the simulation. The daily mean values of precipitation, runoff, evapotranspiration, and groundwater inflow simulated from the watershed models were summed to provide mean annual rates and volumes derived from each year of the simulation.

Mean annual bias for the calibration period for Ash Canyon Creek and Clear Creek watersheds was within 6 and 3 percent, and relative errors were about 18 and -2 percent, respectively. For the 1980–2007 period of record, mean recharge efficiency and runoff efficiency (percentage of precipitation as groundwater inflow and runoff) averaged 7 and 39 percent, respectively, for Ash Canyon Creek, and 8 and 31 percent, respectively, for Clear Creek. For this same period, groundwater inflow volumes averaged about 500 acre-feet for Ash Canyon and 1,200 acre-feet for Clear Creek. The simulation period for the ephemeral watersheds ranged from water years 1978 to 2007. Mean annual simulated precipitation ranged from 6 to 11 inches. Estimates of recharge efficiency for the ephemeral watersheds ranged from 3 percent for Eureka Canyon to 7 percent for Eldorado Canyon. Runoff efficiency ranged from 7 percent for Eureka Canyon and 15 percent at Brunswick Canyon. For the 1978–2007 period, mean annual groundwater inflow volumes ranged from about 40 acre-feet for Eureka Canyon to just under 5,000 acre-feet for Churchill Canyon watershed. Watershed model results indicate significant interannual variability in the volumes of groundwater inflow caused by climate variations. For most of the modeled watersheds, little to no groundwater inflow was simulated for years with less than 8 inches of precipitation, unless those years were preceded by abnormally high precipitation years with significant subsurface storage carryover.

Introduction

Although the rapid population growth witnessed until the late 2000s has abated, the demand for water resources in the Carson River basin (fig. 1) will likely continue to increase. Changes in land and water use and the effects of these changes on groundwater and surface-water resources are uncertain and currently under investigation. In the middle Carson River basin, upstream of Lahontan Reservoir, agricultural land is being urbanized, groundwater pumping is increasing, and changes in surface water and groundwater use (currently for agriculture) will likely cause alterations in groundwater inflow and discharge.

The projected changes may affect flow of the river and, in turn, affect downstream water users dependent on sustained river flows to Lahontan Reservoir. The groundwater and surface-water systems are thought to be well connected in the Carson River basin upstream of Lahontan Dam in Dayton and Churchill Valleys (Brown and Caldwell, 2004; Harrill and Preissler, 1994; Maurer and others, 2009). In these valleys, groundwater pumping may cause the outflow of the Carson River from Lahontan Reservoir to Lahontan Valley to decrease over time. In contrast, the elimination of flood irrigation with diversions from the Carson River and the irrigation of land with treated effluent rather than diversions from the Carson River may cause the flow of the river to increase (Maurer and Berger, 2007, p. 53).

In light of the uncertainties that future land and water use practices may have on the Carson River, an evaluation of groundwater flow and groundwater and surface-water interactions in the middle Carson River basin is needed to provide water managers with information for water-resources planning. The middle Carson River basin includes the hydrographic areas[1] (HA) of Eagle (HA104), Dayton (HA103), and Churchill Valleys (HA102) upstream of Lahontan Dam (fig. 1).

The U.S. Geological Survey (USGS), in cooperation with the Bureau of Reclamation (BOR), began a study in 2008 to develop a numerical model to simulate groundwater and surface-water interactions in the Carson River basin upstream of Lahontan Dam and downstream of Carson Valley. This report describes the development of precipitation-runoff watershed models used to simulate perennial and ephemeral

tributary runoff and groundwater flow (hereafter referred to as groundwater inflow, defined as that component of groundwater recharge derived from deep percolation from excessive rainfall or snowmelt) from the major watersheds within the middle Carson River basin. The USGS Nevada Water Science Center has developed numerous precipitation-runoff models for watersheds in the Sierra Nevada (Jeton and others, 1996; Jeton, 1999a, 1999b, Koczot and others, 2005) and Carson Valley (Jeton and Maurer, 2007). A complementary report to this study (Maurer, 2011) summarizes the hydrogeologic setting and provides the basis for a conceptual model of the groundwater and surface-water systems in the middle Carson River basin, to be developed later in the study.

Purpose and Scope

To assist in evaluating groundwater inflow, this report documents the development and calibration of precipitation-runoff models for 2 watersheds with gaged, perennial streams and 10 watersheds with ungaged, ephemeral streams in the middle Carson River basin for water years (the period between October 1 and September 30 of the following year) encompassing 1978–2007. Model results were compared to measured flow for the perennial watersheds and to mean annual estimates of runoff from the ephemeral watersheds. The data used to develop and calibrate, where possible, the watershed models presented in this report provide estimates of runoff tributary to the middle Carson River and groundwater inflow.

Geographic and Geohydrologic Setting

The middle Carson River basin is in west-central Nevada, extending a distance of about 60 mi and covering an area of about 900 mi[2] (fig. 1). The headwaters of the Carson River lie at altitudes of 10,000 to 11,000 ft in the Sierra Nevada in Alpine County, California. The main stem of the Carson River flows from Carson Valley a few miles southeast of Carson City, the capital of Nevada, into the easternmost part of the Churchill Valley hydrographic area (fig. 2). For purposes of this report, the middle Carson River basin is defined as the area from Eagle Valley on the west downstream to Lahontan Dam.

---

[1]The U.S. Geological Survey and the Nevada Division of Water Resources delineated formal hydrographic areas in Nevada systematically in the late 1960s for scientific and administrative purposes (Cardinalli and others, 1968). The official hydrographic-area names, numbers, and geographic boundaries continue to be used in U.S. Geological Survey scientific reports and Nevada Division of Water Resources administrative proceedings and reports. Hydrographic-area boundaries generally coincide with drainage-area boundaries.

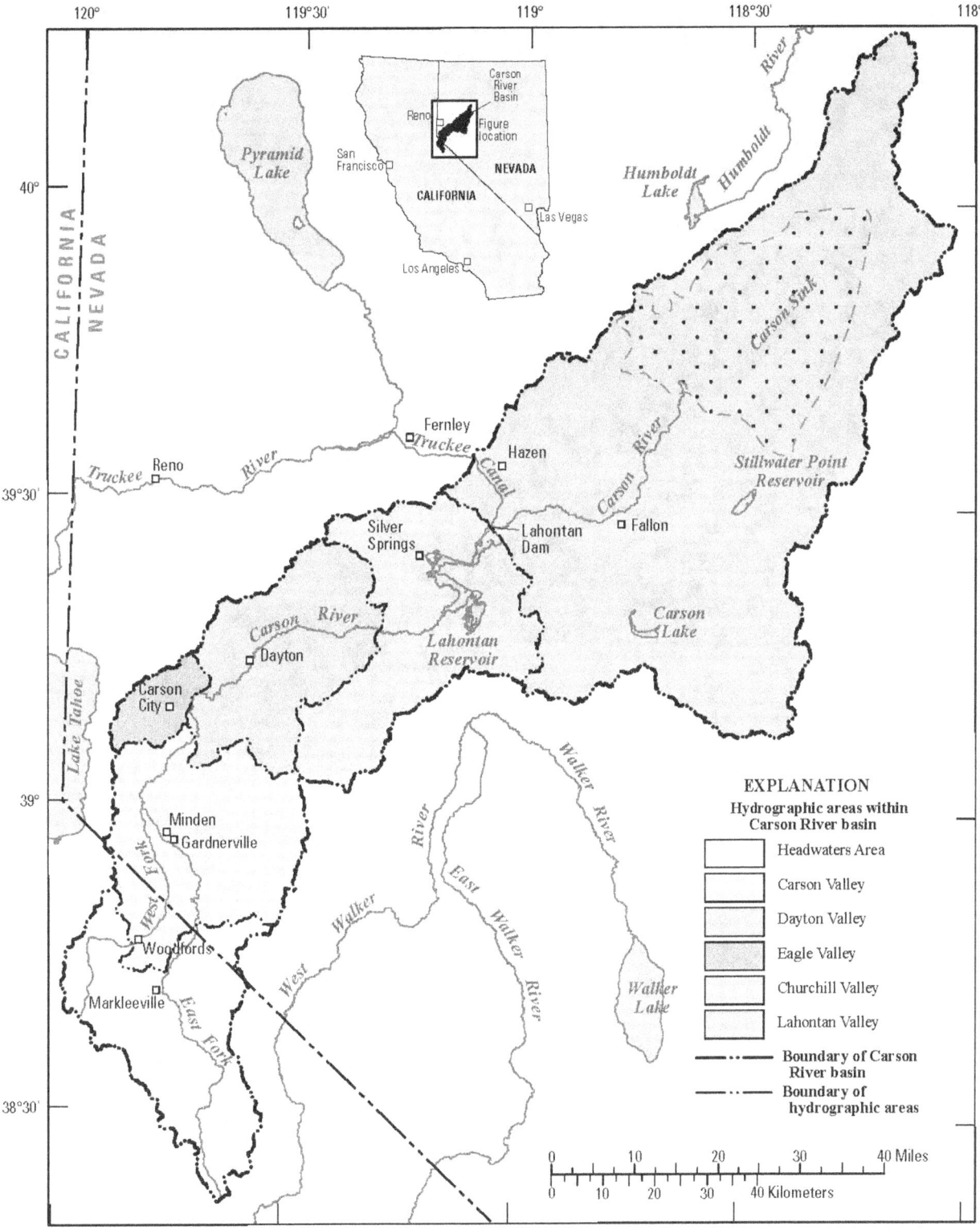

Figure 1. Location of the middle Carson River basin, hydrographic areas within the basin, and selected geographic features, Nevada.

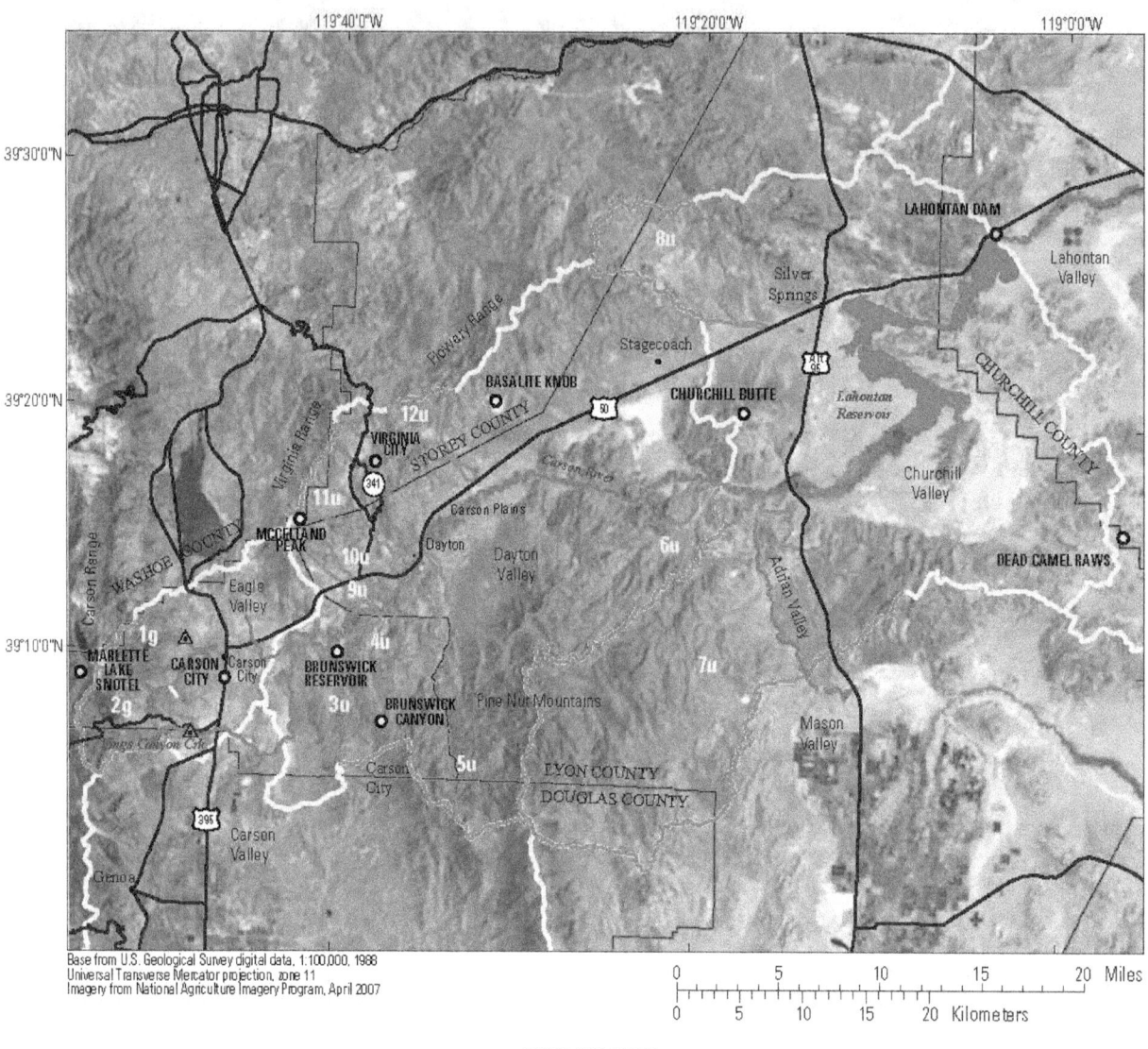

EXPLANATION

▲   Gaging station

○   Precipitation storage gage

◎   Climate station

⎯⎯⎯   Boundary of modeled watershed

         Watershed number (see table 1)

         Boundary of hydrographic areas

Figure 2.   Selected geographic features of the middle Carson River basin, Nevada, and the locations of streamflow gaging and climate stations used in the watershed model development.

Streamflow through the Dayton Valley and Churchill Valley hydrographic areas is diverted for flood irrigation extending about 0.5 mi from the river on the narrow flood plain, which is incised about 50 ft into the floor of both valleys. Adrian Valley drains small wetlands near the northern end of Mason Valley and is tributary to the Carson River southeast of Churchill Butte (fig. 2); however, inflow to the Carson River is largely ephemeral. The river exits the Dayton Valley hydrographic area south of Churchill Butte and enters Lahontan Reservoir about 8 mi downstream of the boundary of the Churchill Valley hydrographic area. Annual streamflow of the Carson River is extremely variable, ranging from a low of about 26,000 acre-ft (or a mean annual streamflow of 35.9 ft^3/s) in 1977 to slightly more than 800,000 acre-ft (or a mean annual streamflow of 1,100 ft^3/s) in 1983 near Fort Churchill (Maurer and others, 2009). In the remainder of this report, the hydrographic areas will be referred to using only the name of the valley they represent.

Most of the Carson River basin lies in the rain shadow of the Sierra Nevada, with precipitation decreasing abruptly from about 38 inches per year (in/yr) at the crest of the Carson Range to about 10 in/yr on the floor of Eagle Valley. Precipitation over the middle Carson River basin ranges from 14 in/yr at Virginia City, Nev., in the western part of the Virginia Range, to only 5 in/yr at Lahontan Dam (obtained at http://www.ncdc.noaa.gov/oa/climate/stationlocator.html, accessed January 30, 2008).

Tributary streamflow to the Carson River is perennial in only three watersheds within the study area and all are located in Eagle Valley (Clear Creek, Ash Canyon and Kings Canyon Creeks); the flow from two, Ash Canyon and Kings Canyon Creeks, has been increasingly diverted for municipal supply to Carson City. As a consequence of the diversions, these watersheds provide streamflow to the river only during spring runoff in excessively wet years when runoff reaches flood stage. Tributary streamflow in the remaining part of the middle Carson River basin is largely ephemeral with flow reaching the Carson River only during excessively wet years.

Extensive discussion on the hydrogeology, geology, and groundwater movement in the study area is detailed in a companion hydrogeologic report (Maurer, 2011), which is briefly summarized here. Groundwater in Eagle Valley generally flows from the west and north toward the center of the valley, then eastward into Dayton Valley where groundwater flows toward the Carson River. In Carson Plains, groundwater flows parallel to the Carson River from southwest to the northeast. Near Stagecoach, groundwater flows from the west, north, and south toward the center of the valley, then northeastward into Churchill Valley. In northern Churchill Valley, groundwater flows eastward from Stagecoach Valley (east of Silver Springs) toward the center of Churchill Valley. In the southern part of the valley, groundwater beneath Churchill Canyon and Adrian Valley flows northward toward the Carson River, then parallel to the Carson River and northeastward toward Lahontan Reservoir. Groundwater levels in the middle Carson River basin show little long-term change from the 1970s to 2007, with the exception of declining water levels in wells on the western side of Eagle Valley near Dayton, on the western side of Carson Plains, and in wells in the Stagecoach area (Maurer and others, 2009). Downward trends in groundwater levels may be a consequence of the combination of increased municipal and agricultural pumping and the effects of dry years from 1999 to 2004.

Maurer and others (2009) used data collected at 14 USGS gaging stations and 22 Federal Water Master gaging stations for various streamflow analyses in the upper Carson River basin, which included the headwaters area in the Sierra Nevada and the hydrogeographic areas upstream of Lahontan Reservoir. Cumulative annual streamflow and associated differences at gaging stations near the boundary of Dayton Valley show an average annual decrease in the flow of Carson River for water years 1940–2006 of about 11,000 acre-ft. The decrease in streamflow is a consequence of the evapotranspiration by the irrigated crops and pasture grasses and infiltration to groundwater storage. The Carson River gains flow through Dayton Valley during or after years of above mean annual streamflow and precipitation. The gains are from tributary inflow and groundwater seepage to the river and occur from 1 to 2 years after above mean annual streamflow and precipitation. Above mean annual streamflow and precipitation for at least 1 year was required to replenish groundwater storage and produce streamflow gains in the following year. Statistical analyses of mean annual streamflow for 1940–2006 showed that the effects of groundwater pumping and changes in land use and water use on Carson River streamflow through the middle Carson River basin were not measurable or were masked by variations in annual precipitation.

Granitic and metamorphic geologic rock units are generally exposed near each other, comprising most of the bedrock surrounding Eagle Valley; the units likely underlie basin-fill sediments that fill the valley. Granitic and metamorphic rocks of the Carson Range are cut by swarms of west-dipping normal faults and the granitic rocks are deeply weathered to depths exceeding 100 ft. The watersheds of Ash Canyon and Clear Creeks primarily consist of weathered granite. Small, scattered exposures of granitic and metamorphic geologic units are found near Adrian Valley, suggesting they may underlie the Tertiary volcanic rock and sedimentary units at relatively shallow depths at that location. The Pine Nut Mountains are composed of several blocks, bounded on the east by north-trending normal faults that have exposed the granitic and metamorphic basement rocks on the east, and tilted the blocks and the mountain range as a whole to the west (Moore, 1969, p. 18).

The basin-fill hydrogeologic unit is in the center of each valley and consists of unconsolidated sediments deposited by streams forming alluvial fans surrounding the valleys, fluvial sediments deposited by the Carson River, and lake sediments deposited during high stands of ancient Lake Lahontan. Lake sediments deposited by ancient Lake Lahontan are present in the basin-fill hydrogeologic unit from the Carson Plains subbasin downstream to Lahontan Valley. Ancient Lake Lahontan covered much of northwestern Nevada at various times during the Pleistocene epoch and its level varied in response to changing glacial climates (Morrison, 1964, p. 110). The sediments deposited by Lake Lahontan vary greatly in lithology. During high lake stands, sediments deposited by the Carson River likely formed deltas in the western part of the Carson Plains subbasin. During low stands when the lake was dry, the Carson River meandered across the valley floors and sand dunes and sand sheets likely covered much of the valley floors as described by Morrison (1964, p. 102–103). As the levels of ancient Lake Lahontan rose and fell, the deposition of deltaic sediments at the mouth of the Carson River moved upstream and downstream, and the deposition of beach and deep-lake clay deposits moved laterally across the valleys, likely creating a complex mixture of Quaternary sediments within the basin-fill hydrogeologic unit. Field observations at the mouth of Bull-Mineral Canyon, one of the ephemeral drainages in the current study, suggest the complex channel morphology evident at this site may be remnant channel terraces from this period.

Description of Watershed Models

Watershed models were developed for 2 gaged perennial-stream watersheds, Ash Canyon Creek (watershed 1g, table 1) and Clear Creek (watershed 2g, table 1), and 10 ungaged ephemeral-stream watersheds in the study area (fig. 2 and table 1). For brevity, in the remainder of the report, the modeled watersheds will be referred to as perennial and ephemeral watersheds, although the term applies to the streams themselves. While there are several gaged watersheds in Eagle Valley, only two, Clear Creek and Ash Canyon Creek, had adequate streamflow records that reflected minimal diversions and outflow to the Carson River. The bedrock geology and hydroclimatology of the Eagle Valley watersheds are somewhat distinct from the middle Carson River basin; however, this upper part of the study area contributes to Carson River flow, particularly during high snowmelt runoff years.

Table 1. Watershed designation, temperature and precipitation index stations used for model development, and period of simulation, middle Carson River basin, Nevada.

Watershed number [1] (location shown in fig. 2)	Watershed name	Period of record used in model simulation (water years)[2]	Temperature index station	Low altitude index precipitation station	High altitude index precipitation station[3]
1g	Ash Canyon Creek	1979–2007	Carson City	Carson City	Marlette Lake
2g	Clear Creek	1990–2007	Carson City	Carson City	Marlette Lake
3u	Brunswick Canyon	1977–2007	Carson City	Carson City	—
4u	Hackett Canyon	1977–2007	Carson City	Carson City	—
5u	Eldorado Canyon	1977–2007	Virginia City	Virginia City	—
6u	Bull–Mineral Canyon	1977–2007	Lahontan Dam	Lahontan Dam	—
7u	Churchill Canyon	1977–2007	Lahontan Dam	Lahontan Dam	—
8u	Ramsey Canyon	1977–2007	Lahontan Dam	Lahontan Dam	—
9u	Eureka Canyon	1977–2007	Carson City	Virginia City	—
10u	Daney Canyon	1977–2007	Carson City	Virginia City	—
11u	Gold Canyon	1977–2007	Carson City	Virginia City	—
12u	Sixmile Canyon	1977–2007	Virginia City	Virginia City	—

[1] Station number ending in "g" is gaged; station number ending in "u" is ungaged.

[2] Includes model initialization year (first year of record used).

[3] "—" under "High altitude index precipitation station" indicates only the lower altitude data was used.

Models for the 10 ephemeral watersheds were initially derived from the ephemeral watershed model representing the east side of Carson Valley, upstream from the present study area (Jeton and Maurer, 2007). Models were developed for ephemeral watersheds to estimate the quantity of ephemeral runoff tributary to the Carson River and the potential for groundwater inflow.

Precipitation-Runoff Modeling System

Conceptually, perennial and ephemeral watersheds such as those in the middle Carson River basin can be described in terms of a few key hydrologic processes that, working in combination, result in measured runoff variations (Beven, 2001). The model used in this study is the Precipitation-Runoff Modeling System (PRMS; Leavesley and others, 1983). PRMS is a process-based, distributed-parameter modeling system designed to analyze the effects of precipitation, climate, and land use on runoff and watershed hydrology (Leavesley and others, 1983). Additional information on PRMS can be found at http://wwwbrr.cr.usgs.gov/projects/ SW_MoWS/software/oui_and_mms_s/prms.shtml.

The term "process-based" refers to the use of mathematical equations to simulate the physical processes of the various water-budget components. The term "distributed-parameter" refers to the representation of the watershed with spatially varying hydrologic characteristics, which is represented numerically as a collection of hydrologic response units (HRUs) that each have a unique set of physical-parameter values. The term "parameter" used throughout this report refers to a numeric constant in equations used to describe hydrologic processes.

In distributed-parameter precipitation-runoff models, the hydrologic processes are parameterized to account for the spatial and temporal variability of basin characteristics. Although partitioning methods differ, the intent of distributed-parameter models is to better conceptualize hydrologic processes, to represent these processes at time and space scales similar to those in nature, and to reduce model input error, thereby improving overall model performance.

As mentioned, the spatial variability of land characteristics that affect runoff within watersheds is accounted for in the model by dividing the modeled area into HRUs. A critical assumption is that the hydrologic response to uniformly distributed precipitation and simulated snowmelt is homogeneous within each HRU. HRUs are thus characterized by those physiographic properties that determine hydrologic response: altitude, slope, aspect, vegetation, soil, geology, and climate. HRUs may consist of noncontiguous or contiguous areas of similar properties. For this study, a 300-meter-square cell representing the regional groundwater model grid was used as the boundary for an HRU, rather than a hydrographic area. Given the coarseness of the

physiographic data, homogeneity within an HRU grid cell was assumed. Water and energy balances reflecting physical and hydrologic characteristics and the climate conditions are computed daily for each HRU. The HRU is indexed to one or more nearby climate stations. Monthly temperature lapse rates and precipitation-correction factors are used to extrapolate measured daily air temperature and precipitation from the nearby climate stations to individual HRUs, thereby accounting for spatial and altitude differences. The form of precipitation (rain, snow, or mixed) is dependent on relations between a specified snow-rain threshold temperature and minimum and maximum temperatures for each HRU.

Responses to climate events can be simulated in terms of water and energy balances, streamflow regimes, flood peaks and volumes, soil-water relations, and groundwater inflow (represented by the term "groundwater sink" in fig. 3). Groundwater inflow from the watersheds moves in the subsurface to become groundwater inflow to the basin-fill aquifers in the middle Carson River basin.

The watershed system is conceptualized as a series of interconnected reservoirs, whose collective output produces the total hydrologic response (fig. 3). The water-budget components (rectangular boxes) denote the storage and collection of water and energy. Daily precipitation, daily maximum and minimum air temperature, and a surrogate for daily solar radiation are inputs that drive the model. The surrogate for solar radiation is estimated from daily temperature using a modified degree-day method and adjusted for slope and aspect and is appropriate for use in the study area because predominantly clear skies prevail on days without precipitation (Frank and Lee, 1966; Swift, 1976). Snowmelt is a significant component of the water budget for mountainous watersheds. Snowpack components of PRMS simulate the initiation, accumulation, and depletion of snow on each HRU. The snowpack is simulated both in terms of its water storage and as a dynamic-heat reservoir (Anderson, 1973; Obled and Rosse, 1977; Leavesley and others, 1983). A snowpack water balance is computed daily within each HRU, and a snowpack energy balance is computed each day and night. For moderate-altitude, snow-dominated watersheds such as in the Virginia Range and the Pine Nut Mountains, the importance of seasonal differences in temperature and precipitation is reflected in snowpack accumulation and melt rates, and ultimately the timing of runoff.

Potential evapotranspiration (PET) was computed using a modified version of the Jensen Haise method (Jensen and Haise, 1963; Jensen and others, 1969). During model calibration, annual PET simulated by PRMS was compared to regional PET values (Farnsworth and others, 1982). In PRMS, PET is first satisfied in the model by vegetation canopy-interception storage, followed by sublimation (snowpack evaporation) and impervious-surface evaporation. When snow is present and there is no transpiration, sublimation is

PRECIPITATION-RUNOFF MODELING SYSTEM

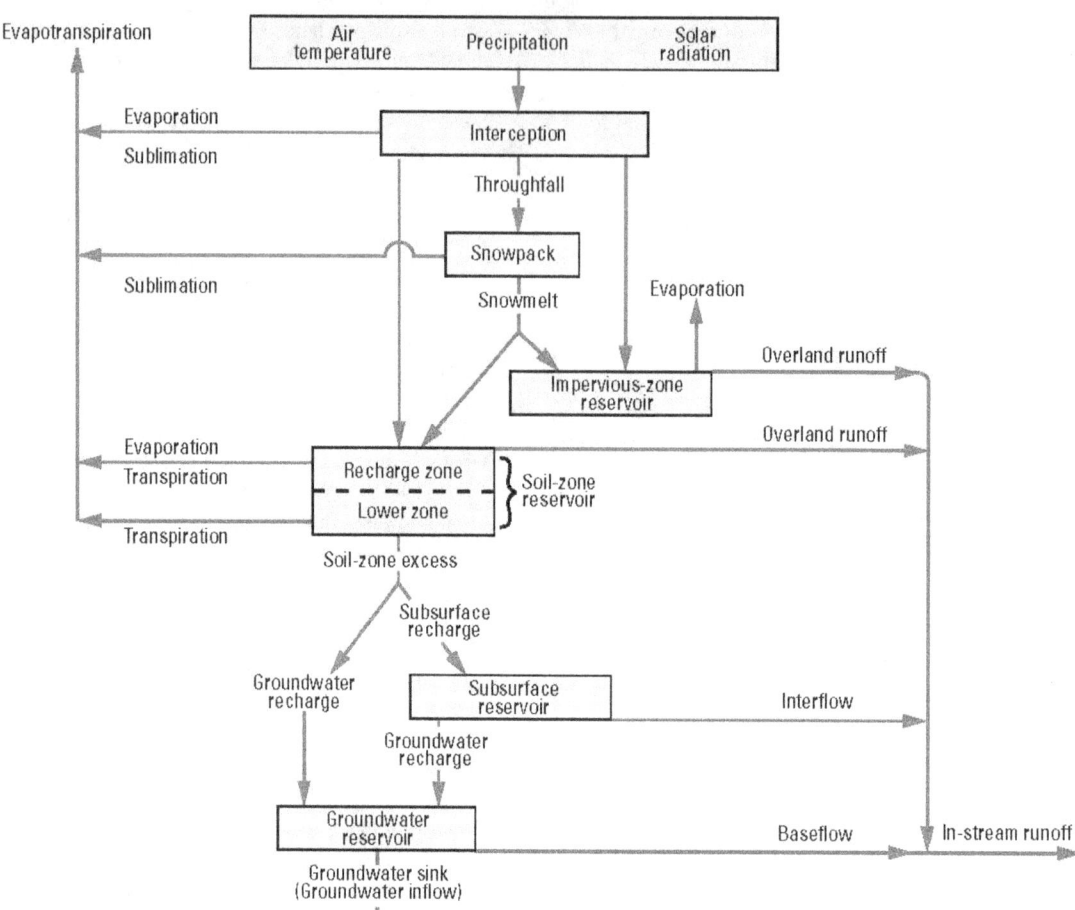

Figure 3.   Schematic of processes simulated by the Precipitation-Runoff Modeling System (Leavesley and others, 1983).

computed as a percentage of the total PET (PRMS assumes no sublimation when plants are transpiring). The remaining PET demand is satisfied by evaporation from the soil surface and soil-zone storage after transpiration begins. The transpiration period depends on the plant type and altitude zone contained within each HRU. For each year of simulation, a cumulative degree-day index is computed (using daily mean temperature) to determine the start of transpiration, allowing for earlier or later initiation of the transpiration period during warmer or cooler springs, respectively.

PRMS simulates the soil zone as a simplified two-layer system: a shallow, upper zone (called the recharge zone in fig. 3) where water losses are from soil evaporation and transpiration, and a deeper, lower zone where the soil-moisture depletion is by transpiration, groundwater and subsurface

recharge. In this study, the subsurface is defined as the unsaturated zone below the root zone and above the water table. The total soil profile depth for each HRU is defined as the average rooting depth of the dominant vegetation. Actual evapotranspiration losses from the soil zone are simulated as proportional to the remaining PET demand and the ratio of currently available soil moisture to the maximum water-holding capacity of the soil profile. In PRMS, infiltration into the soil-zone reservoir depends on the daily snowmelt or net rainfall rates (total precipitation minus canopy interception), soil field capacities, specified maximum infiltration rates (for snowmelt), and antecedent soil-moisture conditions (water in the soil zone prior to infiltration). Infiltration thresholds are defined depending on whether the water is derived from rain or snowmelt.

The subsurface reservoir represents the pathways that the soil-water excess takes in percolating through the shallow unsaturated zones to stream channels, arriving at the streams above the water table. Soil water in excess of field capacity is first used to satisfy recharge to the groundwater reservoir and is assumed to have a maximum daily limit. Once this limit is reached, further percolation of soil water is routed to the subsurface reservoir. Water can then be further allocated to the groundwater reservoir or routed directly to the stream channel from the subsurface reservoir (fig. 3). The latter is referred to as interflow and is computed as a non-linear rate using the storage volume of the reservoir and user-defined routing coefficients. Flow from the groundwater reservoir is the source of baseflow in the stream. Movement of groundwater outside the modeled watershed is simulated by decreasing the groundwater storage and labeling this portion of the water budget as a groundwater sink. In this study, the groundwater-sink flux represents groundwater inflow to the basin-fill aquifers of the middle Carson River.

Runoff, as simulated by PRMS, is a summation of three components: (1) overland runoff from saturated soils or runoff from impervious surfaces, (2) interflow from the unsaturated zone below the root zone as described above, and (3) baseflow. A basic assumption in PRMS is that the runoff travel time, from the headwaters to the outlet of a defined model area (a tributary watershed, for example) is less than or equal to the daily time step, and thus daily runoff need not be explicitly routed along stream channels.

In PRMS, the groundwater reservoir can be thought of as a bucket from which water in storage is released at a rate that fits the baseflow component of the measured hydrograph (the seasonal runoff recessions). Baseflow is designed to respond more slowly to hydrologic fluctuations than interflow. The interflow component typically is represented in the stream hydrograph as the more immediate response to snowmelt, though less rapid than the overland flow component, which occurs when net precipitation or snowmelt exceed infiltration thresholds.

Model Development

The development of the PRMS models required delineating the watershed boundaries of the perennial and ephemeral watersheds, compiling daily time series of runoff, precipitation and minimum and maximum air temperature data, delineating HRUs, and computing initial model parameters. PRMS parameters for the perennial models were derived from similar watersheds in the Carson Valley modeling study (Jeton and Maurer, 2007). Likewise, initial

values for the ephemeral watersheds used parameter values developed for the east side of Carson Valley, considered to be drier than the Carson Range to the west. Similar evapotranspiration rates, soils, vegetation type, and density characteristics apply to the ephemeral watersheds downstream of Eagle Valley. While the HRU-dependent parameters were determined and computed for watershed-specific areas, the non-HRU dependent parameters were transferred from previously calibrated models. Parameters of particular relevance are those used in the routing of water through the soil zones and the shallow subsurface reservoir, precipitation and temperature adjustments, groundwater flow coefficients, and most importantly, those used for simulating groundwater inflow to the basin-fill aquifers.

Basin Characterization and Description of Digital Data

Geographic information system (GIS) software and the Weasel toolbox (Viger, 2008; Viger and Leavesley, 2007) were used to manage spatial data and to characterize model drainages and HRUs in terms of slope, aspect, altitude, vegetation cover densities and types, and soil types, available water-holding capacity, and depths. The GIS Weasel is a software system designed to aid users in determining estimates of spatially varying HRU-specific model parameters as input to lumped and distributed parameter environmental simulation models. For this study, a 30-meter digital elevation model (DEM: U.S. Geological Survey, 1999) was used to delineate the watershed boundaries. Other digital data include slope and aspect (derived from the 30-meter DEM), soils [1:250,000 State Soil Geographic (STATSGO) database (U.S. Department of Agriculture, 1991)], and land cover for computing vegetation type and canopy density. The 30-meter 2001 National Land Cover Data (NLCD) database (http://www.epa.gov/mrlc/nlcd-2001.html) was used to determine the dominant vegetation type, percentage of impervious surface, and vegetation canopy density for each HRU.

The HRUs were delineated as 300- by 300-meter grid cells. Figure 4 shows an example of the gridded HRU delineation for the Ash Canyon Creek watershed in Eagle Valley. The HRU data layer was intersected with digital measurements of altitude, slope, aspect, vegetation, and soils, and averaged values were assigned to each HRU. As the NLCD land cover classification in figure 5 illustrates, the vegetation type within a cell may not be homogeneous; however, at the resolution of the grid cells, most of the HRUs are dominant in one of the five major land cover types.

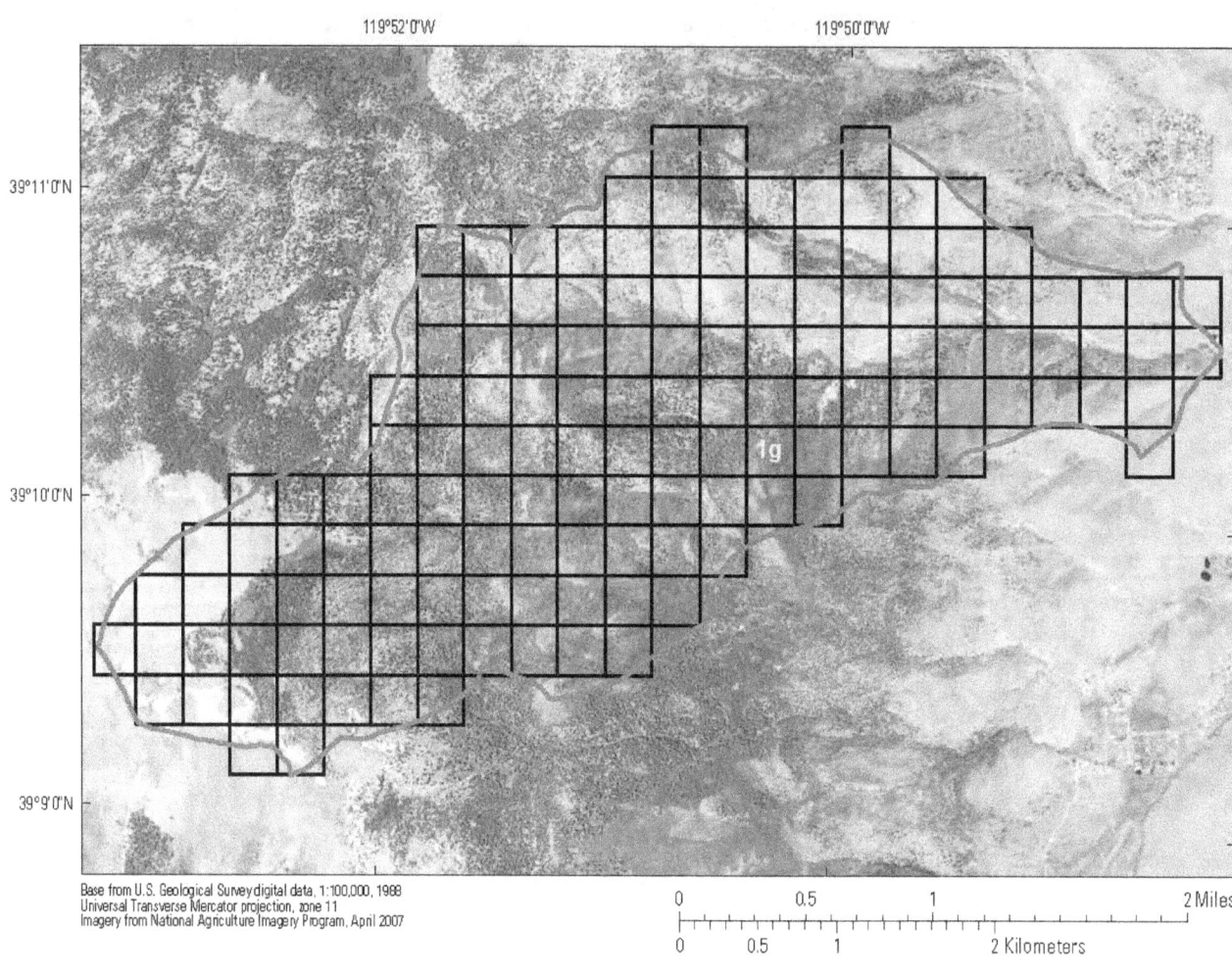

Base from U.S. Geological Survey digital data, 1:100,000, 1988
Universal Transverse Mercator projection, zone 11
Imagery from National Agriculture Imagery Program, April 2007

EXPLANATION

Watershed number (see table 1)
Boundary of modeled watershed
Stream channel
Hydrologic response unit (HRU)

Figure 4.   Gridded Hydrologic Response Unit delineation for Ash Canyon Creek, Eagle Valley, middle Carson River basin, Nevada.

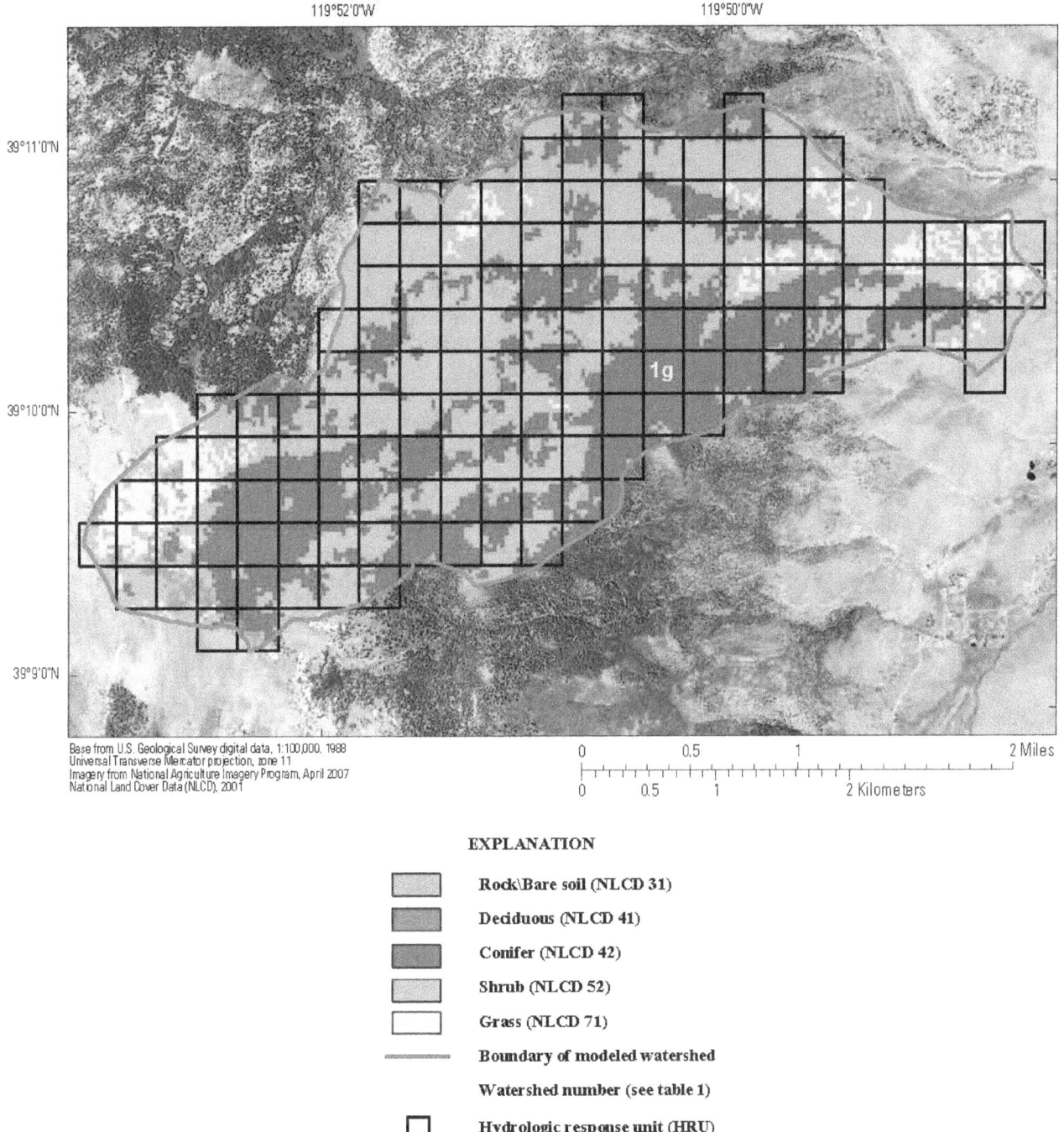

Base from U.S. Geological Survey digital data, 1:100,000, 1988
Universal Transverse Mercator projection, zone 11
Imagery from National Agriculture Imagery Program, April 2007
National Land Cover Data (NLCD), 2001

EXPLANATION

Rock\Bare soil (NLCD 31)

Deciduous (NLCD 41)

Conifer (NLCD 42)

Shrub (NLCD 52)

Grass (NLCD 71)

Boundary of modeled watershed

Watershed number (see table 1)

Hydrologic response unit (HRU)

Figure 5.   National Land Cover Data for Ash Canyon Creek, Eagle Valley, middle Carson River basin, Nevada.

When comparing the NLCD vegetation classification to the high-resolution aerial imagery (fig. 4), the spatial distribution of land cover is well represented in the NLCD data. The GIS-derived parameters are "static," meaning they are simulated as constant through time and are not adjusted during model calibration. Typically, watershed models are run using several years of daily climate data as model input, and land cover and density are assumed to be constant over time. In the present study, the land cover data reflect conditions from 1998 to 2000. Whereas vegetation cover type and canopy density for many areas within the middle Carson River basin underwent changes attributed to wildfires, urban development, and drought prior to 1998, changes have commonly been related to urbanization since 2000. Sublimation (evaporation from snow) and the longwave net radiation component of the simulated energy balance are partially affected by HRU vegetation canopy densities, which in turn affect snowmelt and runoff timing more than overall runoff volume. While using a static land cover data set can be problematic when concerned about daily hydrologic simulations, accurate simulation of runoff timing was less of a concern in the present study because the water-budget components were aggregated to annual and mean annual values.

Rapid growth and associated urbanization of Eagle Valley and Dayton Valley since the mid-1990s has increased the extent of impervious surface area with significantly more impervious cover mapped in the NLCD 2001 data set than was present in the earlier part of the modeling record. This is particular the case for Daney, Eureka, and Sixmile Canyon watersheds. However, for the remaining modeled watersheds, impervious area is mainly limited to roads. Initial global model parameters, whose values apply over the entire basin, were quantified from PRMS parameter values for similar watershed studies in the region (Jeton and others, 1996; Jeton, 1999a and 1999b, Jeton and Maurer, 2007). The perennial and ephemeral watersheds are hydrographically defined basins, defined as land areas that drain to a downstream point.

Point precipitation and temperature measurements from climate stations at lower or higher altitudes than the HRUs are distributed to the HRU using orographic corrections based on the mean HRU altitude. Using a gridded cell framework for HRU delineation allows for less spatial variability in altitude within a particular HRU than were the HRU defined by drainage networks and thousand-foot altitude zones, as in previous studies. Restricting the range in altitude within a single HRU decreases the magnitude of the orographic corrections.

Runoff and Climate Data

Ash Canyon Creek and Clear Creek are perennial streams and flow onto the floor of Eagle Valley during most years. Flow remaining after municipal and agricultural diversion exits Eagle Valley (and for Clear Creek, flows into the north end of Carson Valley), and discharges into the Carson River. Ash Canyon Creek and Clear Creek have continuous streamflow gaging records dating back to 1976 and 1989, respectively. The USGS records indicate daily flow to Ash Canyon Creek may be influenced by diversions from Marlette Lake and Hobart Creek Reservoir, while flow to Clear Creek may also be affected by a small-scale diversion upstream of the gage (Garcia and others, 2002). Maurer and Berger (1997) noted the potential for groundwater flow from the Clear Creek watershed across the hydrographic divide toward Carson Valley. Neither measured record was modified to account for upstream diversions nor out-of-basin groundwater flow.

The term "measured" runoff is used for measured or continuously gaged runoff, and "simulated" runoff is defined as runoff simulated by the watershed model. With the exception of a few indirect measurements (streamflow measurements estimated from geomorphic evidence such as high-water marks, channel scouring, or change in channel geometry typically determined in the absence of streamflow measurements) of fair to poor quality for individual runoff events, no measured runoff data exists for any of the 10 ephemeral watersheds. Instead, mean annual runoff estimates derived from regional runoff estimates for each modeled ephemeral watershed (Moore, 1968) were used as a coarse comparison to the period of record simulated. Moore determined mean runoff per unit area for successive altitude zones based on estimates of mean flow derived from streamflow measurements, or where none existed, measurements of channel cross sections. Because Moore cautions against using these estimates where local variations in geology, precipitation, vegetation, and land-use may vary as the runoff-altitude relations represent much larger areas than the modeled watersheds, these estimates are used here as a general comparison to mean annual simulated runoff.

Climate input-data requirements for PRMS are daily total precipitation and daily maximum and minimum air temperature. Daily precipitation from four stations in and near the middle Carson River basin (stations 1p–4p, table 2; figs. 2 and 6) were used to determine daily precipitation for each HRU in each watershed model (Western Regional Climate Center, 2008). The climate record used for modeling spanned a period of 30 years, from water year 1978 to 2007.

Table 2. Climate and streamflow–gaging stations, period of record, and altitude, middle Carson River basin, Nevada.

[Station number ending in "p" is precipitation stations, station ending in"sp" are storage gage sites, station number ending in "g" are streamflow gages. Station locations are shown in Figure 2. Altitude: Datum is North American Vertical Datum of 1988. Abbreviations; USGS, U.S. Geological Survey; BLM, Bureau of Land Management; SNOTEL, snowpack telemetry; RAWS, Remote Automated Weather Stations]

Station identifier	Station name (Site identifier)	Source	Period of record used in model simulation (water years)[1]	Station altitude (feet)
		Precipitation stations		
1p	Carson City (#261485)	NWS[2]	1977–2007	4,651
2p	Marlette Lake SNOTEL (#19K04S)	NRCS[3]	1979–2007	7,880
3p	Virginia City (#268761)	NWS	1977–2007	6,340
4p	Lahontan Dam (#264349)	NWS	1977–2007	4,130
5sp	McClellan Peak (#391532119420601)	USGS	1997–2007	7,410
6sp	Brunswick Canyon (#390726119371901)	USGS	1997–2007	6,370
7sp	Brunswick Reservoir (#391011119395201)	USGS	1997–2004	5,100
8sp[4]	Basalite Knob (#392037119312201)	USGS	1997–2007	5,580
9sp	Churchill Butte (#392024119173901)	USGS	1997–2007	6,004
10sp	Dead Camel RAWS	BLM	1997–2007	4,490
		USGS Streamflow gages		
11g	Ash Canyon Creek (#10311200)		1979–2007	5,080
12g	Clear Creek (#10310500)		1990–2007	5,000

[1] Period of record used in watershed model simulation includes the first year of record used for model initialization.

[2] National Weather Service Cooperative Station.

[3] Natural Resources Conservation Service SNOTEL station.

[4] Storage precipitation gage.

The stations used for each watershed model initially were selected by their geographic proximity to the watershed and the altitude distribution within the watershed. Higher altitude, continuously recording climate data were limited to the SNOwpack TELemetry (SNOTEL) data (U.S. Department of Agriculture, 2008) from the climate station at Marlette Lake (7,880 ft). The Virginia City site (6,340 ft), located about two thousand feet higher than the valley floor, was considered a mid-altitude station. Two other sites, Carson City (4,651 ft) and Lahontan Dam (4,150 ft) at the western and eastern extent, respectively, of the study area, reflect precipitation amounts for the valley floor. Eagle Valley, and to a greater extent the Dayton Valley and Churchill Valley hydrographic areas, lie in the rain shadow of the Carson Range. Mean annual precipitation at Marlette Lake SNOTEL station averages about 35 in. for the period of record while precipitation at the Carson City gage averages about 10 in. annually. Further east, mean annual precipitation at Virginia City averages 13 in. while the Lahontan Dam gage, located at the eastern edge of the study area, averages only 5 in.

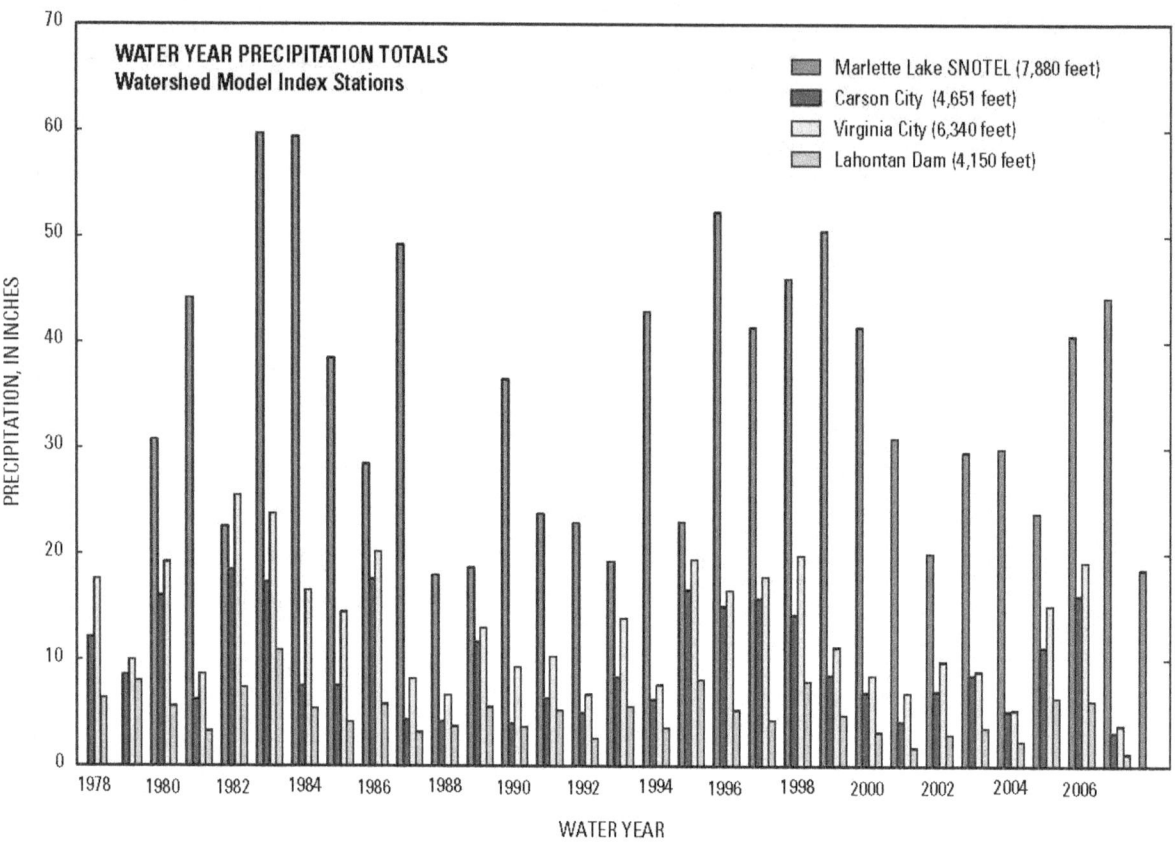

Figure 6.   Annual precipitation at the four index stations used in watershed modeling; Marlette Lake SNOTEL, Carson City, Virginia City, and Lahontan Dam, Nevada, water years 1978–2007.

Data from several storage precipitation gages (which measure volume of precipitation rather than continuous, daily measurements) also were used to evaluate simulated annual precipitation for watersheds within the Dayton Valley and Churchill Valley hydrographic areas (figs. 2 and 7, table 2) that otherwise have no measured climate data. While these data are for a shorter record (water years 1997–2007), this period represents a sample distribution of high and low precipitation years found within the modeling period. The storage gages at McClellan Peak, Brunswick Canyon, Brunswick Reservoir, Basalite Knob, and Churchill Butte represent otherwise ungaged areas and provided some measure of annual precipitation distribution. The Dead Camel site (4,490 ft) near Lahontan Dam is a continuous-recording Remote Automated Weather Station (RAWS) with a shorter period of record than the four stations used in modeling. For this report, however, it is considered as ancillary data along with the storage gages mentioned. Mean annual totals for the storage gages ranged from a high of 9.6 in. for Brunswick Canyon to 4.2 in. at the Dead Camel site. The McClellan Peak storage gage located near and at about 1,000 ft higher than the Virginia City gage recorded only an average of 9 in. for the 1997–2007 period versus the 11 inches recorded at the Virginia City gage for this

same period. Mean annual precipitation as plotted in figure 7 with decreasing station altitudes illustrates the effect of decreasing precipitation from west to east, particularly when comparing the Marlette Lake gage to the McClellan Peak gage, a difference of only 400 ft in altitude.

Precipitation data for the Churchill Canyon watershed was recorded for an earlier period of record (1964–1980) using storage precipitation gages for 16 locations spanning altitudes from 4,620 ft to near 7,000 ft (Joung and others, 1983). Mean annual precipitation ranged from 7 to 13 in. Precipitation totals for one high-altitude station (near the western upper boundary (6,960 ft) and one low-altitude station (4,390 ft) near the mouth of Adrian Valley (not shown) indicates the largest difference between the two sites includes five of the eight driest years (1966, 1968, 1972, 1974, and 1977). This suggests less precipitation at the lower altitude sites may be due to a more pronounced rain shadow effect for dry years.

Initial PRMS model simulations for the Eagle Valley perennial watersheds (Ash Canyon Creek and Clear Creek) used an HRU precipitation correction factor that increased precipitation 15 to 20 percent for each 1,000 ft of altitude gain above the valley floor. This initial correction factor was derived from local lapse rates calculated using the Carson

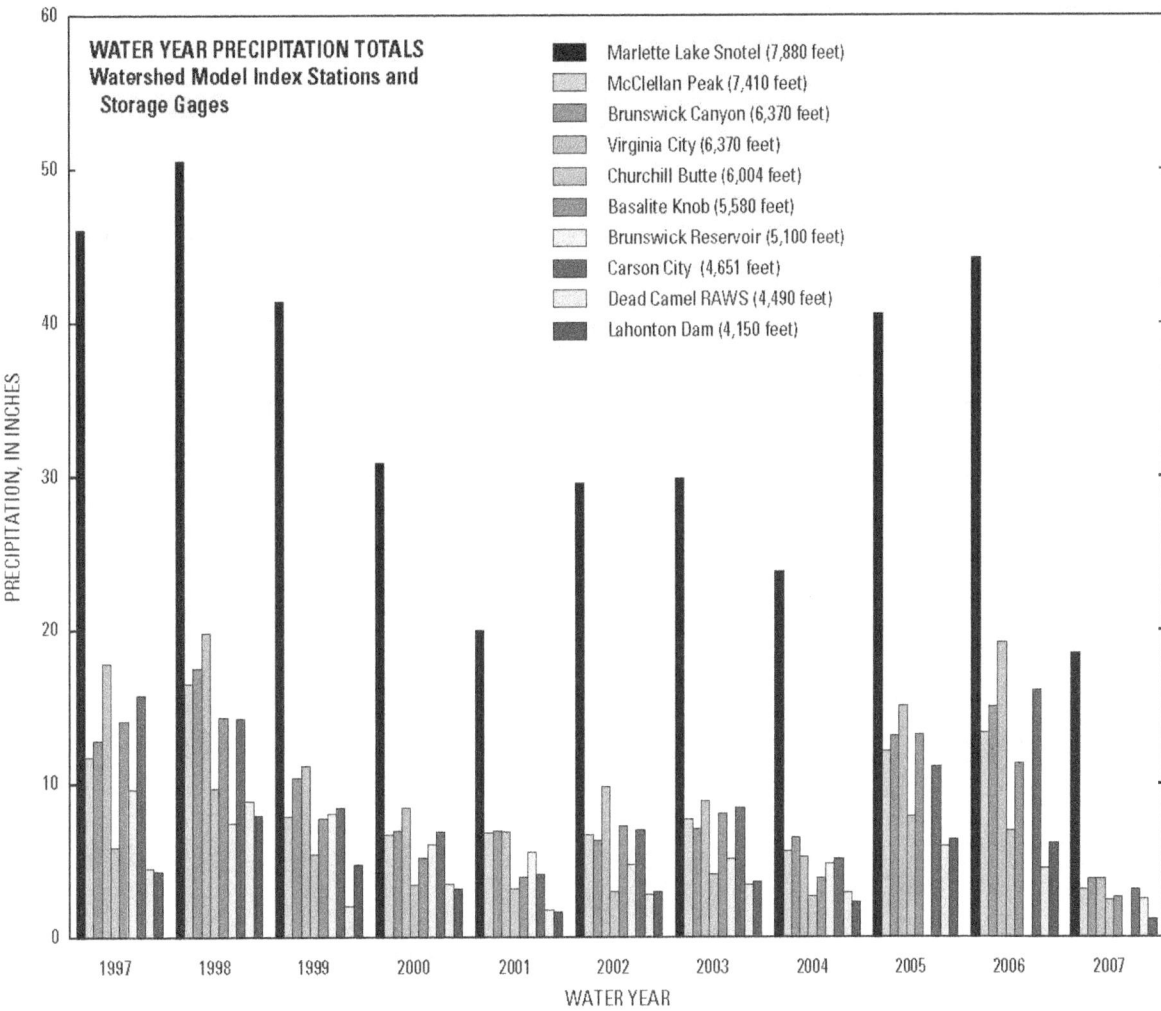

Figure 7. Annual precipitation for five storage gages and five continuously recording gages, ordered in decreasing elevation, middle Carson River basin, Nevada, water years 1997–2007.

City station as the low-altitude precipitation station and the Marlette Lake SNOTEL as the high-altitude precipitation station and differences in mean HRU altitude.

For the 10 ephemeral drainages east of Eagle Valley, there were no two continuously recording low- and high-altitude stations from which to compute initial precipitation lapse rates. Though the Carson City climate station is influenced locally by the rain shadow of the Carson Range, precipitation amounts continue to decrease significantly east of Eagle Valley. For this reason (with the exception of Brunswick and Hackett Canyon watersheds) the Carson City station was not used in modeling precipitation for the watersheds downstream of Carson City. The rain shadow effect is evident when comparing the Marlette Lake SNOTEL gage to the Virginia City gage (fig. 6). The altitude difference between the two stations is just over 1,000 ft, yet the Marlette site records on average 63 percent more annual precipitation.

Initial adjustments used a ratio of mean annual precipitation derived initially from a statistical distribution of precipitation using correlation and regression analyses (Lopes and Medina, 2007, and later revised by Maurer and others, 2009), and the period of record mean for the index station used to drive the daily watershed model. Patterns in spatial distribution of precipitation known as the precipitation-zone method (PZM) were identified by Lopes and Medina (2007) by mapping station locations and plotting precipitation normals versus station altitudes. Stations with a similar slope and intercept were grouped into a common geographic zone. Regional areas (termed zones) where precipitation was assumed to be linearly related to altitude were determined for much of northwestern Nevada including the middle Carson River basin. Gridded (raster) estimates of mean annual precipitation were determined using a GIS, 30-meter digital elevation model (DEM) and one of the four regional regression equations developed for west-central Nevada.

The period 1971–2000 was chosen for statistical calculations for comparison to the 30-year normals calculated by several federal agencies. This period is considered to represent a long-term average and the period with the most precipitation data, particularly for the higher altitude data represented by the SNOTEL network.

Maurer and others (2009) adopted this approach (hereafter referred to as the adjusted-PZM precipitation data) using the 1971–2005 period of record and included several high-altitude SNOTEL sites to develop a similar gridded precipitation data set based on the Lopes and Medina (2007) precipitation zones. In the current study, the resulting gridded data set from Maurer and others (2009) was combined with the HRU areas for each modeled watershed and a ratio of the mean annual precipitation estimates for 1971–2005 to the 1978–2007 mean of the index climate station were used to

initially adjust the HRU precipitation correction. Subsequent adjustments to this correction factor were made during model calibration.

Maximum and minimum daily temperatures were adjusted in the PRMS model with an altitude correction factor of 3.5 to 4.5 °F (depending on the month) of cooling for every 1,000 ft of altitude gain, which corresponds to regional temperature lapse rates used in similar watershed modeling studies (Jeton and Maurer, 2007). Table 1 lists the temperature station associated with each modeled watershed. Daily air temperature time series (fig. 8) vary little from year to year within a station record with occasional extremes noted for the minimum temperatures. This consistency from year to year allows for more confidence when using monthly, regional lapse rates mentioned above when adjusting the individual daily HRU temperature.

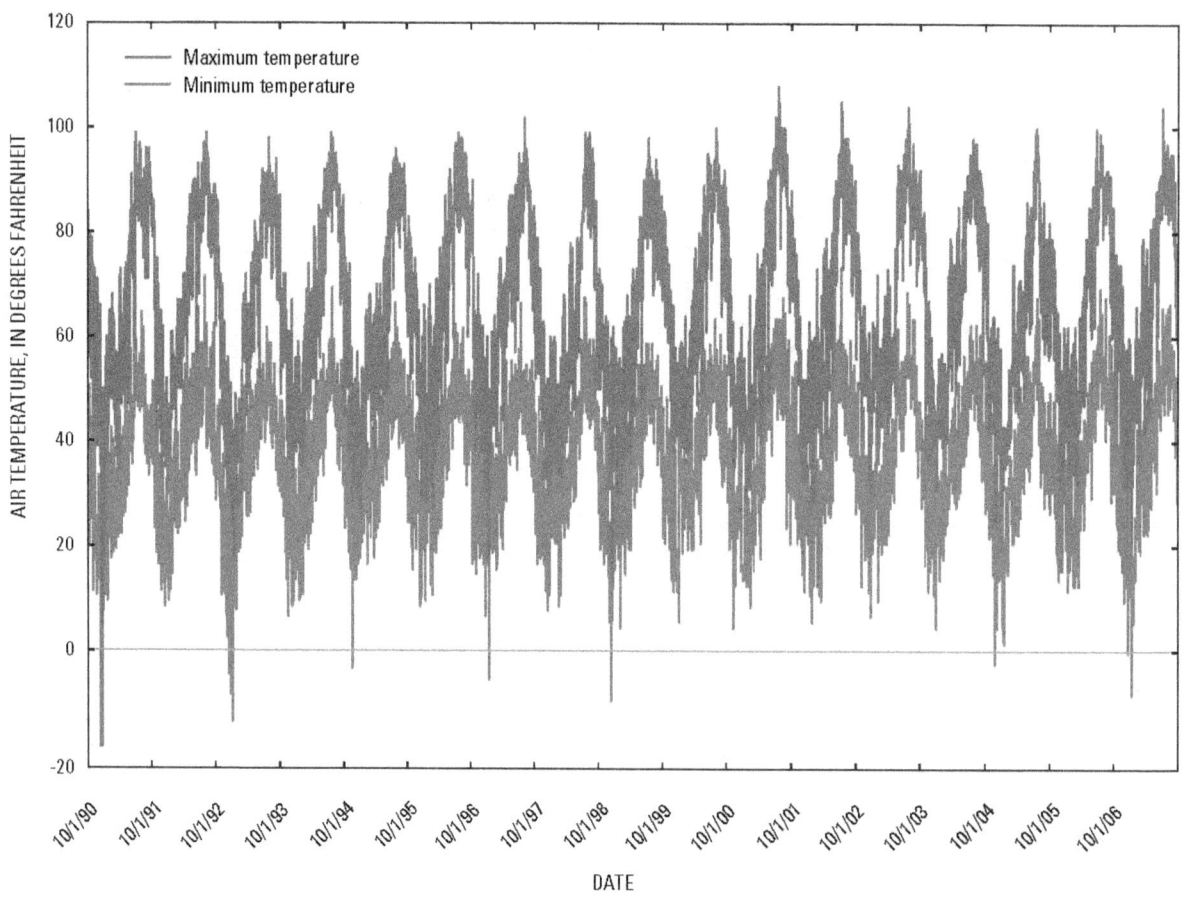

Figure 8.   Daily mean maximum and minimum air temperature recorded at the Carson City station, Eagle Valley, Nevada, water years 1991–2007.

Model Sensitivity

Sensitivity analyses during model calibration typically help to determine the extent to which parameter-value uncertainties result in acceptable runoff predictions. The model sensitivities to PRMS parameter values for the present study can be understood from previous watershed modeling studies in the East Fork Carson River basin (Jeton and others, 1996), the Lake Tahoe basin (Jeton, 1999a), the catchment area of the Truckee River (Jeton, 1999b), and more recently, the Carson Valley area (Jeton and Maurer, 2007). Although this modeling study was focused on estimating groundwater inflow, the hydrologic data to which the watershed model is generally calibrated to is runoff, with groundwater inflow simulated as water in the groundwater reservoir in excess of what reaches the stream channel as baseflow. While only 2 of the 12 modeled watersheds have continuously measured streamflow data, runoff for the ephemeral watersheds was adjusted to better represent the intermittent nature of runoff in these ephemeral drainages while maintaining a reasonable groundwater inflow rate as determined from previous studies.

The hydroclimatic setting of the earlier Carson River basin study (Jeton and Maurer, 2007) is similar to that of the watersheds in the present study area with appropriate adjustments made for precipitation distribution. Previous studies of similar watersheds list the parameters modified during calibration (Jeton, 1999b, p. 17). Prior sensitivity analyses show that runoff simulations are most sensitive to the (1) snow threshold temperature that determines precipitation form, (2) precipitation-correction factor for snow and rain (similar to a precipitation lapse rate where the measured precipitation is adjusted for differences in altitude between the climate station and the HRU), (3) monthly temperature lapse rates (typically between 3.5 and 4.5°F for every 1,000 ft), (4) monthly evapotranspiration coefficients for the Jensen-Haise potential-evapotranspiration computation (Jensen and Haise, 1963), and (5) coefficient for transmission of solar radiation through winter plant canopies to snow surface, which affects snowmelt timing.

The watershed models in this and previous studies also were sensitive to soil moisture storage, and the flow-routing coefficients for interflow and groundwater storage used to simulate groundwater inflow and baseflow. Parameters that determine flows to and from the groundwater reservoirs were adjusted to fit the observed shapes of the seasonal recession of runoff. In the absence of baseflow data, the subsurface parameters were adjusted to fit long-term mean estimates of groundwater inflow from previous studies, and intermittent (or ephemeral) runoff.

Model Calibration

Calibration of PRMS models is an iterative process where, after each adjustment of model parameters, simulated runoff is visually and statistically compared with measured or reconstructed runoff, with special attention (in this study) paid to matching flow volumes for seasonal and annual time periods. If the dominant gains to the system (precipitation) and losses (evapotranspiration) are adequately modeled, and the simulated hydrograph matches the measured hydrograph overall, water in excess of that which reaches the stream channel can be considered as an adequate representation of groundwater inflow. The simulations are run on a daily time step; however, groundwater inflow is evaluated on a mean annual basis to allow for comparison to previously derived estimates (for the Eagle Valley, Maurer and others, 1996; Maurer and Berger, 1997). Seasonal and annual water-budget components derived from the models were of most interest and the detailed timing of runoff and groundwater inflow was not crucial.

Lacking measured streamflow data, calibration of the ephemeral watershed models was limited to the adjustment of precipitation volumes and then comparing simulated runoff volumes to Moore's (1968) mean annual runoff estimates. A finer time-step than annual is unreasonable because the amount of ephemeral runoff is uncertain. An effort was made during calibration to provide the best fit to measured runoff during the spring snowmelt runoff periods for water years 1980 to 2007 for Ash Canyon Creek and 1991 to 2007 for Clear Creek. The first year of simulation was used as an initialization period and thus discounted when computing calibration statistics (water year 1979 and 1990 for Ash and Clear Creek, respectively). For the ephemeral watersheds, the simulation period included water years 1978 to 2007.

Model Simulation Results

No single calibration of a PRMS model will simulate all runoff regimes with equal accuracy. The goal in modeling is threefold: (1) little to no bias, (2) small simulation error, and (3) realistic parameter values reflecting the conditions being modeled. The goals for calibration are to maintain a good visual fit between the simulated and measured hydrographs, to keep mean annual biases to within 5 percent, and to keep relative error to within 10 percent. In watershed modeling, common measures of simulation error include the sum of errors and bias. Bias is computed to determine the presence of systematic error or an indication of central tendency (that is, whether the simulations show a tendency towards under- or overestimating with respect to the measured runoff). Absolute errors (defined as the difference between simulated and measured runoff) tend to be dominated by a few large events (Haan and others, 1982), unless normalized by the measured values to form "relative error," as used in this report. The un-normalized root mean square error (RMSE) provides a common measure of the magnitude of simulation errors that complements the relative measures provided by the bias and relative errors.

Normalizing runoff error by dividing it by the measured value presents a problem when the extremely low flows result in very large relative errors even though the absolute error may be small (Haan and others, 1982). Though much of the measured runoff into Eagle Valley (including runoff from watersheds not discussed in this report) represents very low flows, no runoff data from the gaged watersheds were omitted in the error analysis. Model calibration biases, relative errors, and RMSEs for the two perennial watersheds are given in table 3. Error statistics were not calculated for the ephemeral models because of the considerable uncertainty estimating mean annual runoff. The Moore-derived estimates are presented as a general comparison (table 4); however, the results were not used to validate one model over another. To simulate ephemeral flow, the baseflow component of runoff was adjusted to simulate zero flow during most of the summer months.

The error statistics for the two perennial watersheds are presented as seasonal, mean monthly, and mean annual summaries for the simulation period. Monthly error statistics were computed for four seasons: October–December, January–March, April–June, and July–September. Each of these seasons represents a particular hydroclimatic regime. October–December is characterized by continued baseflow conditions in October and November, with a slight increase in flow due to cessation of evapotranspiration, and climatically a variable period of cold and warm storm fronts producing early winter rain, snow, and mixed rain and snow events. January–March is characterized generally as a snowpack accumulation period though historically rain-on-snow events have produced flood-stage runoff. The spring snowmelt runoff period from April–June produces the most water available for groundwater inflow to basin-fill deposits. July–September is characterized by low-flow or baseflow conditions and occasional runoff from intermittent convective storms.

Table 3.   Calibration statistics for PRMS watershed models for Ash Canyon Creek and Clear Creek, Eagle Valley, middle Carson River Basin, Nevada.

Season [1]	Ash Canyon Creek[1] water years 1980-2007			Clear Creek Water years 1991-2007		
	Bias [2] (percent)	Relative error[4] (percent)	RMSE [3] (inches)	Bias (percent)	Relative error (percent)	RMSE (inches)
October–December	1.5	11.2	0.64	-33.8	-30.5	0.40
January–March	2.4	19.7	0.73	4.3	-2.6	0.70
April–June	5.6	22.3	1.04	5.8	3.7	0.58
July–September	2.9	40.1	0.89	25.6	36.1	0.30
Period of record mean monthly	6.1	6.2	0.08	1.4	1.4	0.09
Period of record mean annual	6.5	18.1	2.53	3.1	–2.1	1.53

[1]Period of record of streamflow data used in computing the PRMS statistics for Ash Canyon corresponds to start date for Marlette Lake SNOTEL climate data (water year 1979) minus the first year for model initialization. The Clear Creek modeling period reflects the start of the gaging record minus the first year for model initialization.

[2]Bias = $\sum$ (simulated – observed)/$\sum$(observed)*100.

[3]Relative error = $\sum$((simulated–observed)/observed)*100)/number of observations.

[4]RMSE is root mean square error = SQRT($\sum$(simulated–observed)2/number of observations).

Table 4. Summary of model results for 2 perennial and 10 ephemeral watersheds in the Middle Carson River Basin, for water years 1978–2007 and 1980–2007, respectively, and comparison with Moore's mean annual runoff estimate (Moore, 1968).

[Location of watershed numbers are shown in figure 2. Watershed number: Station number ending in "g" is gaged; station number ending in "u" is ungaged. Moore's runoff estimate: Mean annual runoff estimate assume no appreciable runoff for basin area less than 5,000 feet in altitude (Moore, 1968). **Mean recharge efficiency:** Groundwater inflow is defined as that component of groundwater recharge derived from deep percolation from rainfall or snowmelt to the basin-fill aquifers of the middle Carson River basin. **Mean runoff efficiency:** Period of record used for computing statistics for Ash Canyon Creek and Clear Creek (water years 1980–2007) is dependent on the start date (10/1/1978) for the Marlette Lake SNOTEL, the high-altitude station used in both models (allowing for 1 year initialization period). **Simulated mean annual groundwater inflow:** Mean efficiencies for recharge and runoff are computed as mean annual groundwater inflow and runoff as a percentage of mean annual precipitation, respectively. Water budget residual: Water budget residual is defined as unaccounted water when (surface runoff + evapotranspiration + groundwater inflow) are subtracted from precipitation. The remaining balance (from column 12) is divided by mean annual precipitation for a percent residual]

Watershed number	Watershed name	Drainage area (acres, rounded)	Simulated mean annual runoff acre-feet (inches)	Moore's runoff estimate acre-feet (inches)	Unit of runoff (acre-feet/acre) using PRMS (column 4)	Mean recharge efficiency (percent)	Mean runoff efficiency (percent)	Simulated mean annual Precipitation acre-feet (inches)	Evapotranspiration acre-feet (inches)	Groundwater inflow acre-feet (inches)	Water budget residual inch	percent
					Perennial watersheds							
1g	Ash Canyon Creek	3,270	2,920 (10.7)	1,850 (6.8)	0.89	7	39	7,410 (27.2)	3,900 (14.3)	490 (1.8)	0.34	1.2
2g	Clear Creek	9,900	4,370 (5.3)	4,040 (4.9)	0.44	8	31	13,940 (16.9)	8,420 (10.2)	1,160 (1.4)	0.04	0.2
					Ephemeral watersheds							
3u	Brunswick Canyon	8,270	1,170 (1.7)	1,100 (1.6)	0.14	4	15	7,930 (11.5)	6,480 (9.4)	340 (0.5)	-0.01	-0.9
4u	Hackett Canyon	4,580	420 (1.1)	230 (0.6)	0.09	6	14	3,090 (8.1)	2,440 (6.4)	190 (0.5)	0	0.0
5u	Eldorado Canyon	35,830	4,480 (1.5)	3,880 (1.3)	0.13	8	13	33,740 (11.3)	26,870 (9.0)	2,690 (0.9)	0	0.0
6u	Bull-Mineral Canyon	18,460	460 (0.3)	1,230 (0.8)	0.02	5	3	14,460 (9.4)	12,310 (8.0)	770 (0.5)	0.58	6.1
7u	Churchill Canyon	98,250	8,190 (1.0)	5,730 (0.7)	0.08	7	11	75,330 (9.2)	62,230 (7.6)	4,910 (0.6)	-0.01	-0.1
8u	Ramsey Canyon	15,340	640 (0.5)	640 (0.5)	0.04	5	8	7,930 (6.2)	6,900 (5.4)	380 (0.3)	0	0.0
9u	Eureka Canyon	2,250	110 (0.6)	40 (0.2)	0.05	3	7	1,500 (8.0)	1,350 (7.2)	40 (0.2)	0.01	0.1
10u	Daney Canyon	4,980	370 (0.9)	120 (0.3)	0.07	4	12	3,110 (7.5)	2,660 (6.4)	120 (0.3)	0	0.0
11u	Gold Canyon	9,100	530 (0.7)	680 (0.9)	0.06	4	9	6,070 (8.0)	5,230 (6.9)	230 (0.3)	0.01	0.1
12u	Six Mile Canyon	8,560	860 (1.2)	570 (0.8)	0.10	5	13	6,850 (9.6)	5,560 (7.8)	360 (0.5)	0	0.0

Watershed modeling results for perennial and ephemeral watersheds are summarized in table 4 and are presented as mean annual estimates for water years 1978 to 2007 for the ephemeral watersheds, and for water years 1980–2007 for the perennial models. The perennial models were limited by the Marlette Lake SNOTEL gage, which began in water year 1979. Water years 1977 and 1979 were used to simulate initial conditions for the ephemeral and perennial watersheds, respectively, and as such are not included in the statistical results. The mean annual runoff estimates from Moore (1968), as described earlier, are presented in table 4 as a gross comparison to the simulated mean annual runoff in lieu of other comparative data. Runoff efficiency is a ratio of runoff to precipitation that indirectly is a measure of losses to evapotranspiration and infiltration. Unit of runoff, runoff per acre, is provided as a comparison of runoff irrespective of drainage area. Recharge efficiency is the ratio of groundwater inflow to precipitation, with both efficiencies computed as a percentage of total precipitation. The water budget residual provides an indication of water not accounted for directly in any of the hydrologic components listed in table 4, and to some degree rounding discrepancies. For example, at the culmination of a model run some water resides in storage (defined as the summation of snowpack water equivalent, groundwater and subsurface reservoir storage, vegetation canopy, and soil moisture storage). Storage calculated at the end of a time step was not accounted for in the water budget residual since this value represents an instantaneous value rather than a mean value, as are the other water budget components listed in table 4. Residual values greater than 1 percent of mean annual precipitation suggest the likelihood that some subsurface storage was unaccounted for. Plots of annual water budget components (precipitation, evapotranspiration, runoff, and groundwater inflow) are provided for the 2 perennial watershed models and 3 of the ephemeral models as representative of 10 ephemeral watersheds. The term "simulated" is implied when discussing water budget components illustrated in the figures presented below.

Perennial Watershed Model Results

The basin-fill aquifer in Eagle Valley is partially recharged by subsurface flow from tributaries draining the Carson Range mountain block. Ash Canyon Creek and Clear Creek (1g–2g, table 1; fig. 2) are considered to be representative perennial watersheds for Eagle Valley. Pathways of surface runoff from the mountain block to the Carson River are obscured by urban development and associated urban drainage networks. The present study only provides estimates of groundwater inflow and surface runoff at the mouth of the modeled watersheds with no attempt to quantify either surface or subsurface flow to the main stem of the Carson River.

Ash Canyon Creek Watershed

Calibration results are summarized from water year 1980 to 2007. Overall bias for the Ash Canyon Creek model was satisfactory for the seasonal aggregates and was only slightly higher for mean monthly and mean annual runoff (table 3). The seasonal relative error, however, overestimated runoff for all seasons from about 11 percent for the October–December season to 40 percent for the summer (July–September) season. Noticeable in the daily hydrograph (fig. 9) for water years 1994 to 1999 is a tendency to overestimate runoff during dry years (represented by water year 1994) while for the wet years (1995–1997) spring runoff was adequately simulated. However, when looking at the full modeling period, the January–February flood events of water years 1986, 1997, and most recently 2005 were undersimulated by as much as 80 percent of the measured mean daily value (fig. 10).

Accurate simulation of warm rain-on-snow events of the magnitude represented in the current streamflow record is possible at the expense of maintaining an adequate winter snowpack for simulating spring runoff. Setting the snowpack threshold temperatures and air temperature warm enough to simulate a rain event from the lower to higher altitudes often resulted both in an earlier spring melt than what was observed and less overall snowmelt runoff. Since these temperature-related parameters were constant over the modeling period, attempting to match the warm flood peaks resulted in less overall simulated snow. Baseflow (the contribution of streamflow from groundwater) varies throughout the measured record from less than 1 ft^3/s during periods of extended below-average precipitation (for example from 1987 to 1994) to above 5 ft^3/s following above-average spring runoff. As noted during the summer months for water year 1994 (fig. 9) and throughout the record during low-precipitation years, measured baseflow dips to an annual low in late summer and increases later into the fall months when little precipitation typically occurs. Simulated baseflow, however, maintained a constant rate, suggesting that more actual evapotranspiration was occurring in the summer than was being modeled (thus increasing baseflow when evapotranspiration shuts downs later in the season), or there are unaccounted diversions upstream from the gage.

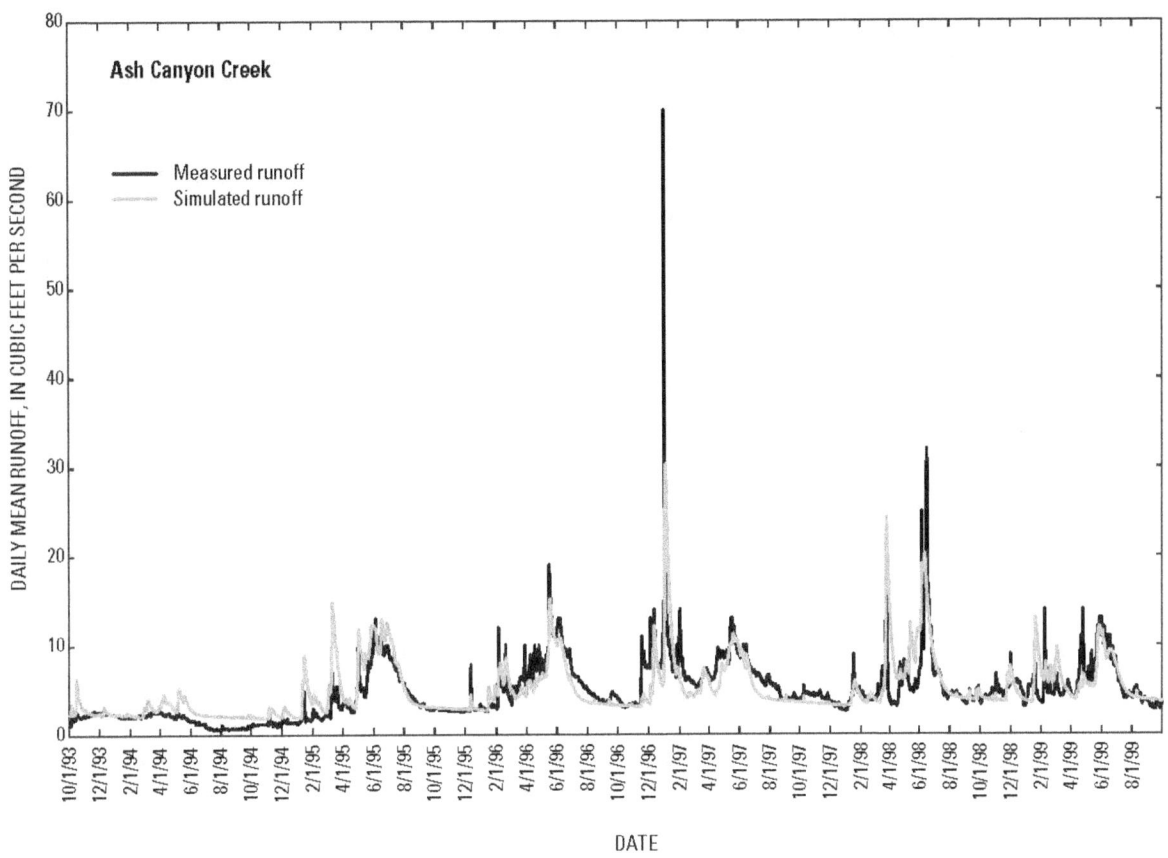

Figure 9.   Simulated and measured daily mean runoff for Ash Canyon Creek, Eagle Valley, Nevada, water years 1994–1999.

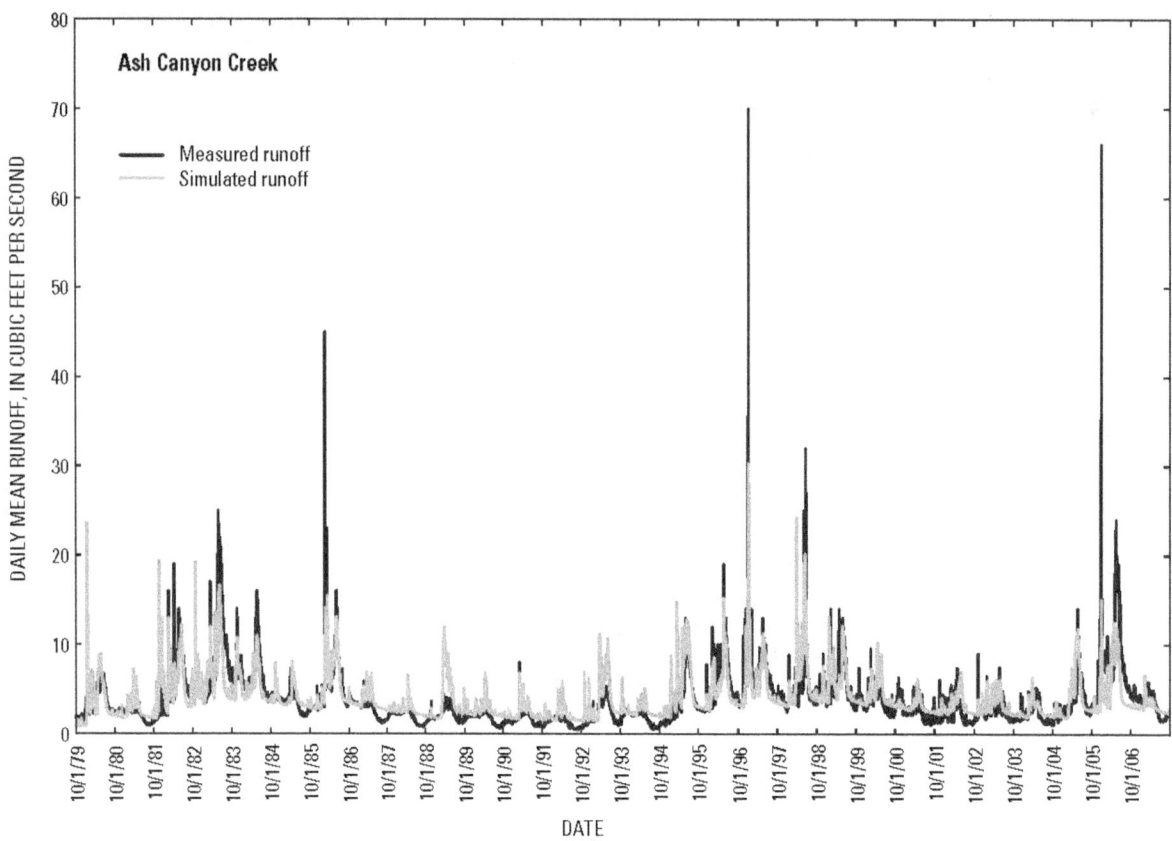

Figure 10.   Simulated and measured daily mean runoff for Ash Canyon Creek, Eagle Valley, Nevada, water years 1980–2007.

Mean monthly flow, computed as a percentage of annual flow (fig. 11) allows for a monthly comparison of simulated and measured runoff relative to total annual runoff. The relative error computed for the seasonal aggregates in table 3, when viewed in the context of the percentage of total annual flow, reduces the influence of relative error. The monthly distribution provides a more realistic estimate volumetrically while providing overall patterns of seasonal runoff distribution. Daily relative error differences over 100 percent are not uncommon during the July to September low flow period (some of this difference is attributed to agricultural withdrawals); however, the difference in percentage of annual streamflow for this season is less than 4 percent. Likewise, the difference for the spring runoff months, which produce on average 35 percent of the annual flow, is less than 1 percent.

Comparisons of simulated and measured annual runoff, groundwater inflow, precipitation, and mean precipitation for the period of record (blue line) are presented in figure 12. Mean annual precipitation (considered here as above 27 in/yr) define above (wet) and below (dry) mean precipitation years in the context of simulation results discussed below. For the Ash Canyon Creek model, on average the relative difference

for measured and simulated annual runoff was 18 percent of measured runoff (table 3). For dry years the model consistently overestimates runoff while for the wet years there is no systematic pattern; the model either closely matches measured runoff (1980, 1986, 1999, 2005) or equally over or underestimates runoff for the other wet years. Annual recharge efficiency (representing groundwater inflow) ranged from 3 to 15 percent of annual precipitation with a mean of 7 percent for the period of record (table 4).

Groundwater inflow varies little from year to year with small increases during wet years primarily as a response to increased snowmelt infiltration as illustrated in figure 12. Mean runoff efficiency for the 1980–2007 period is around 40 percent, slightly higher than earlier estimates, though differences may be attributed to the period of record used and the source of precipitation data. The water budget residual (water remaining once evapotranspiration, groundwater inflow, and runoff are subtracted from precipitation) for Ash Canyon Creek is about 1 percent of the total precipitation. Ideally this number should be zero if all water is accounted for; however, this may represent residual subsurface storage or withdrawals upstream from the gaging station.

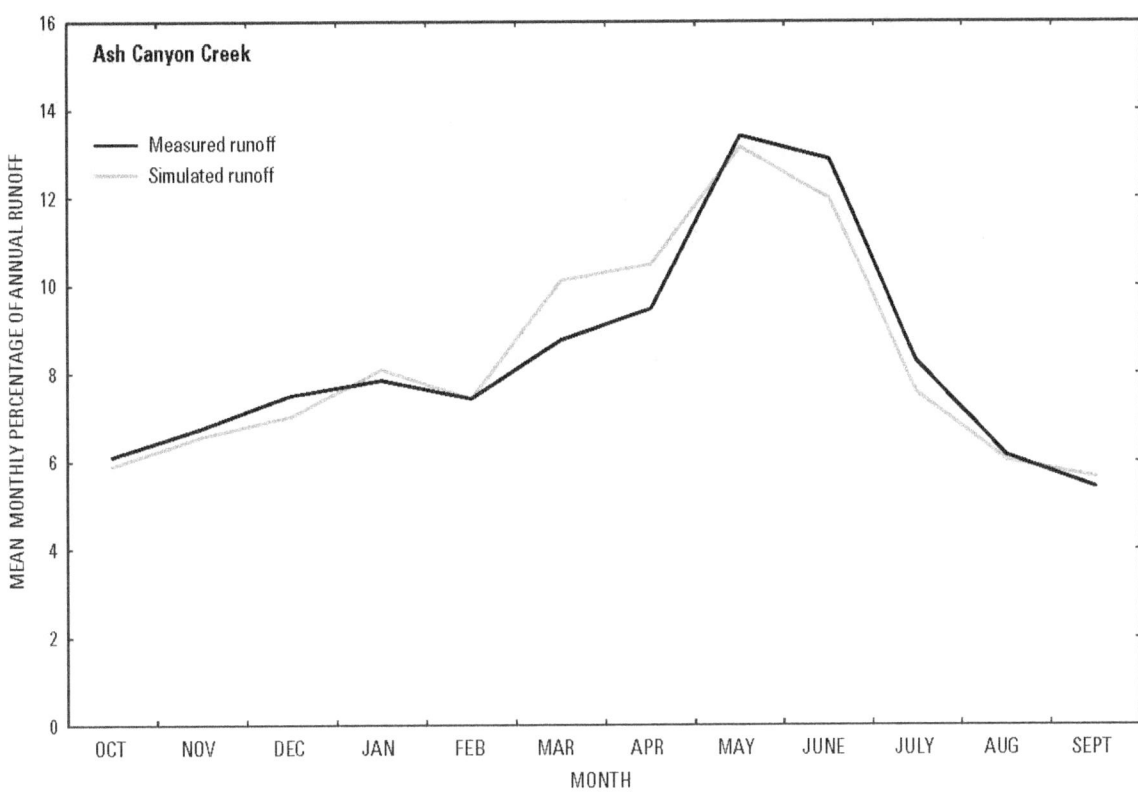

Figure 11. Mean monthly runoff for Ash Canyon Creek, Eagle Valley, Nevada, as a percentage of annual runoff for data collected for water years 1980–2007.

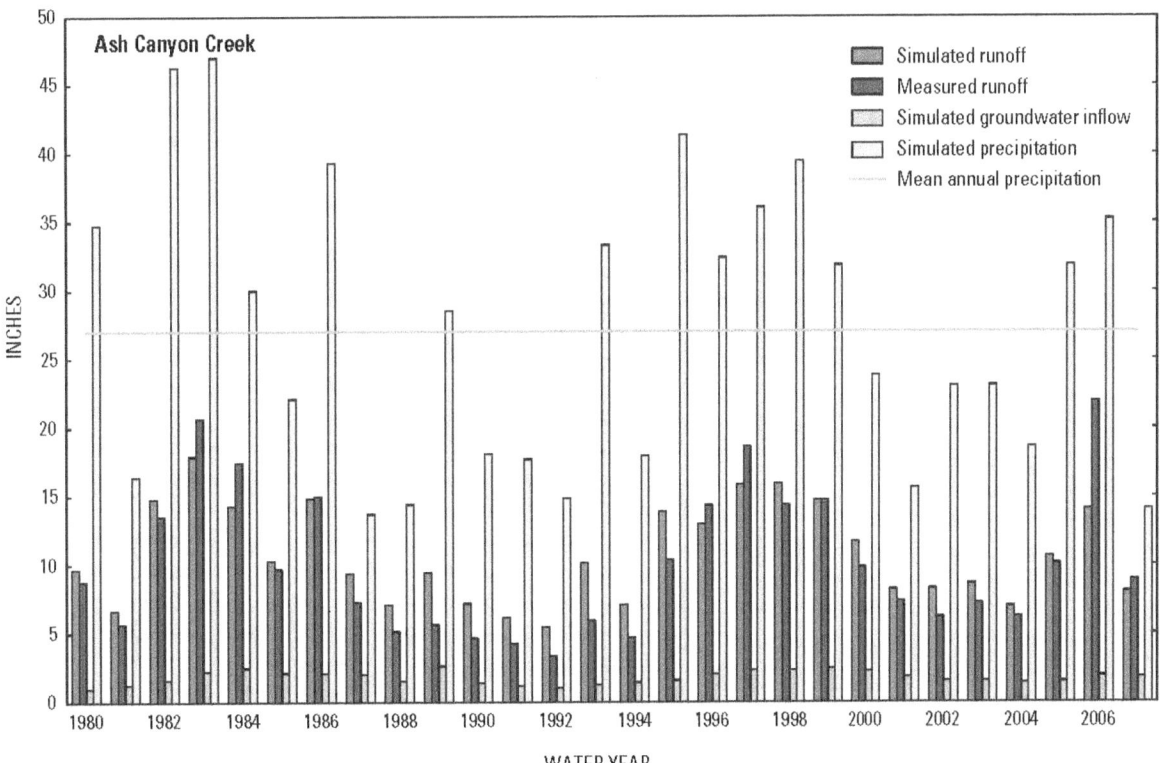

Figure 12. Simulated and measured annual runoff, simulated groundwater and precipitation, and mean annual precipitation (blue line) for Ash Canyon Creek, Eagle Valley, Nevada, water years 1980–2007.

Clear Creek Watershed

The Clear Creek model was calibrated for water years 1991 to 2007. Calibration results for Clear Creek (table 3) indicate acceptable bias and relative error for both mean monthly and mean annual statistics. The seasonal aggregates are reasonable for the winter (January–March) and spring (April–June) periods, which account for roughly 70 percent of the average annual streamflow. In contrast, the model accounts poorly for both the summer (July–September) and fall–early winter months (October–December). Runoff was overestimated by more than 30 percent relative error for the summer period and simulation results indicate a very high bias and relative error towards underestimating the fall period.

Actual fall air temperatures may on average be warmer than simulated, particularly for the higher altitude HRUs, resulting in less precipitation as rain being simulated and less runoff for the fall season. Measured streamflow during both summer and fall months is typically less than 5 ft³/s, hence differences in simulated versus measured runoff produce disproportionately high error relative to the magnitude of flow.

The daily mean hydrograph (fig. 13) for the modeling period from water year 1991 to 2007 indicates overall a better representation of the spring runoff period and snowmelt recession although the model underestimated runoff for three of the wettest years as illustrated for water years 1995, 1997, and 2006. As in the Ash Canyon Creek model, the

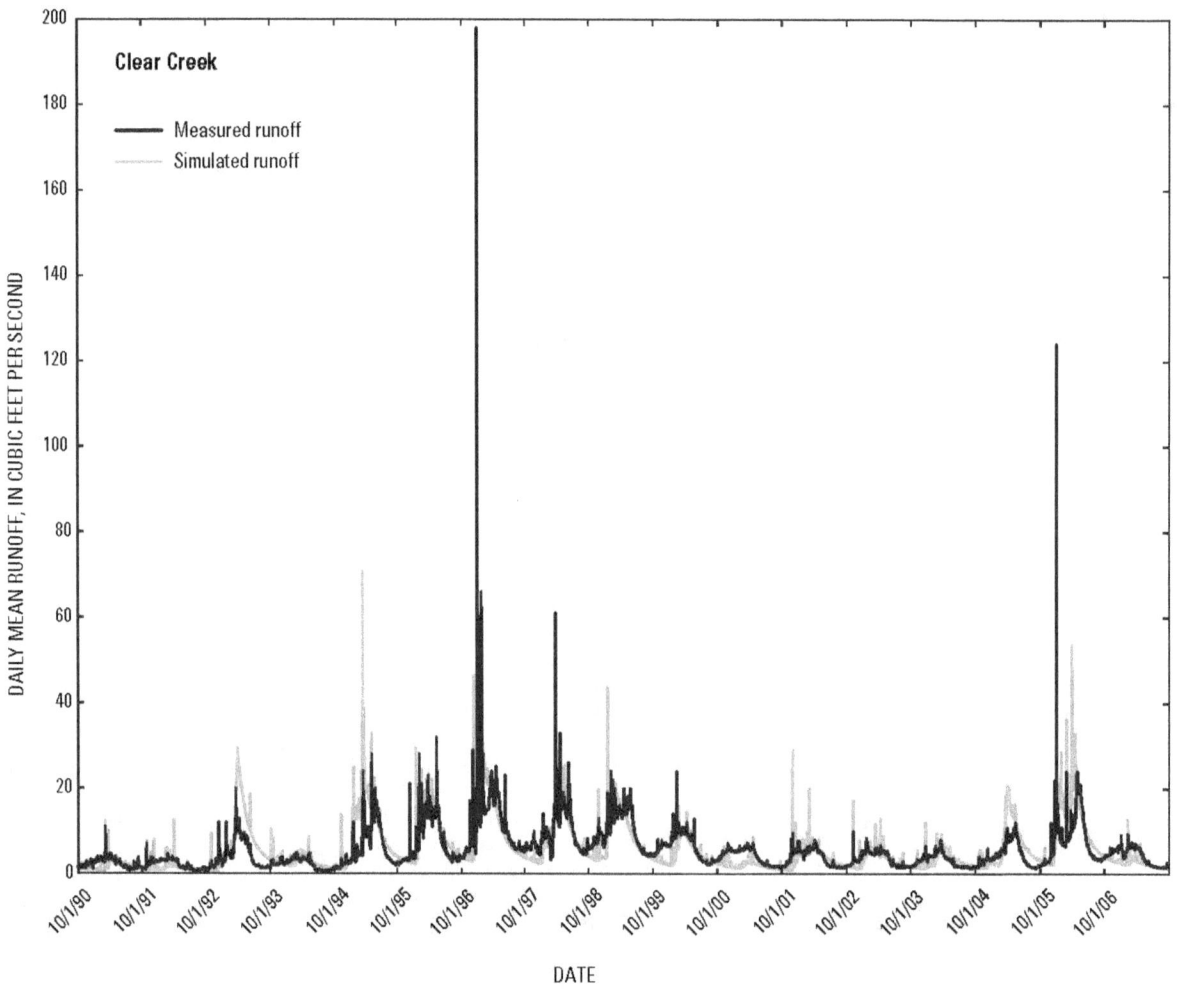

Figure 13.   Simulated and measured daily mean runoff for Clear Creek, Eagle Valley, Nevada, water years 1991–2007.

peak flow events of water years 1997 and 2006 were not well simulated for the reasons described earlier; notably, abnormally warm storms producing precipitation as rain to altitudes well above the normal snow line, rapidly melting the existing snowpack, and producing runoff over much of the watershed. In addition, while frontal in nature, winter storm precipitation amounts varied within Eagle Valley as evidenced by the variable snowpack depth distribution (anecdotal evidence). This suggests the presence of higher intensity cells within the general storm front depositing differing amounts of snow irrespective of altitude. While this may be a common climatic feature for this region, the PRMS models simulate precipitation amount and form (based on temperature) relative to static lapse rates or correction factors based primarily on altitude. Increasing precipitation to match the more pronounced flood peaks resulted in overestimating runoff throughout the period of record. Similarly, decreasing the temperature lapse rates to simulate warmer than average winter storms to improve the flood runoff peaks resulted in a systematic earlier-than-observed depletion of the higher altitude snowpack for much of the modeling record. Agricultural diversions upstream from the gage may account for much of the summer discrepancy, while for the fall months the cessation of evapotranspiration may account for some of the return flow not captured in the model (fig. 14), particularly for riparian vegetation that was not well represented in the model.

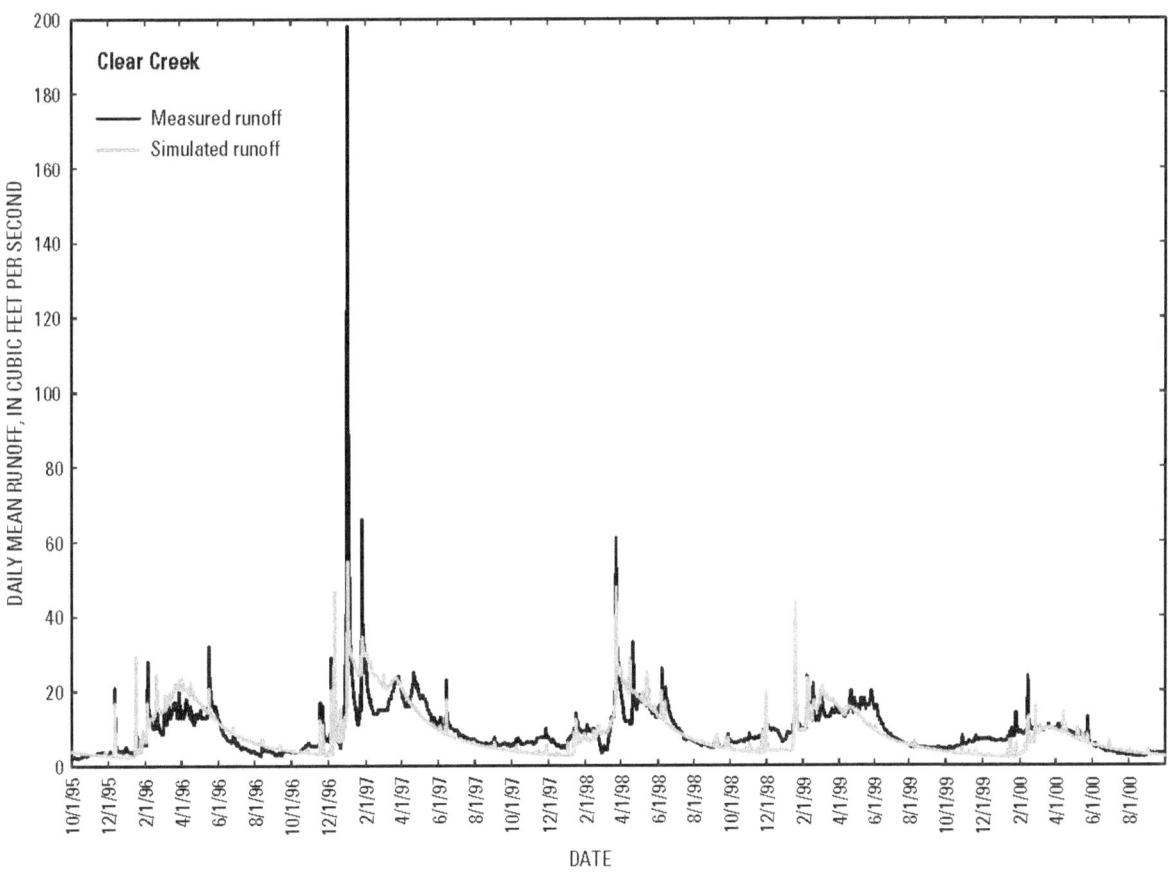

Figure 14. Simulated and measured daily mean runoff for Clear Creek, Eagle Valley, Nevada, water years 1996–2000.

Mean monthly percentage of annual streamflow as shown in figure 15 illustrates a more realistic difference in simulated and measured monthly flows than the error reported for the seasonal aggregates listed in table 3. The March–April differences are attributed to an underrepresentation of November through January runoff, whereby more of the spring snowpack was retained in the model than on the watershed, resulting in an overestimation of runoff during the spring months. Matching the snowmelt recession curve during the wetter periods resulted in an oversimulation of summer flows for drier years, which may also be due to unaccounted for upstream diversions for agricultural use and groundwater outflow to the neighboring Carson Valley (Maurer and Berger, 1997). On average the summer and the fall periods individually produce less than 15 percent of the total annual streamflow (fig. 15). In contrast the winter and spring runoff summed aggregates are within 2 percent of the measured distribution.

The water budget residual for Clear Creek (table 4) is less than 1 percent of mean annual precipitation. This suggests that a recharge efficiency of 8 percent, while higher than the 5 percent estimated by Maurer and Berger (1997, p. 32), may account for some projected subsurface outflow from the Clear Creek watershed into neighboring Carson Valley, in addition to groundwater recharge from Clear Creek itself. Annual recharge efficiency ranged from 5 percent in 1991 to 16 percent in 2007. Higher recharge efficiency for a dry year is not atypical following a wet year as seen for 1994, 2000, and 2007 (fig. 16). This variability is a result of applying a constant groundwater inflow rate over varying subsurface storage. Dry years following wet years can yield higher groundwater inflow than expected due to the previous year's subsurface storage. Runoff efficiency for Clear Creek was about 30 percent for the period of record.

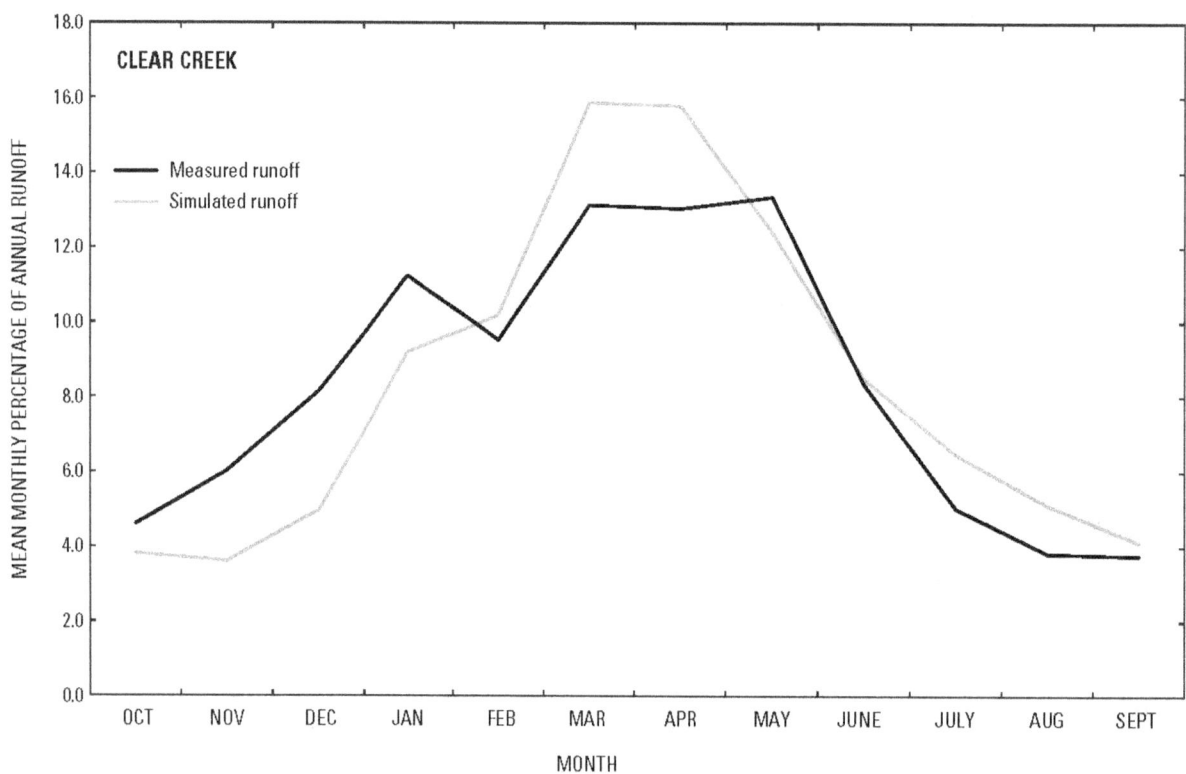

Figure 15.   Mean monthly runoff for Clear Creek, Eagle Valley, Nevada, as a percentage of annual runoff for data collected for water years 1991–2007.

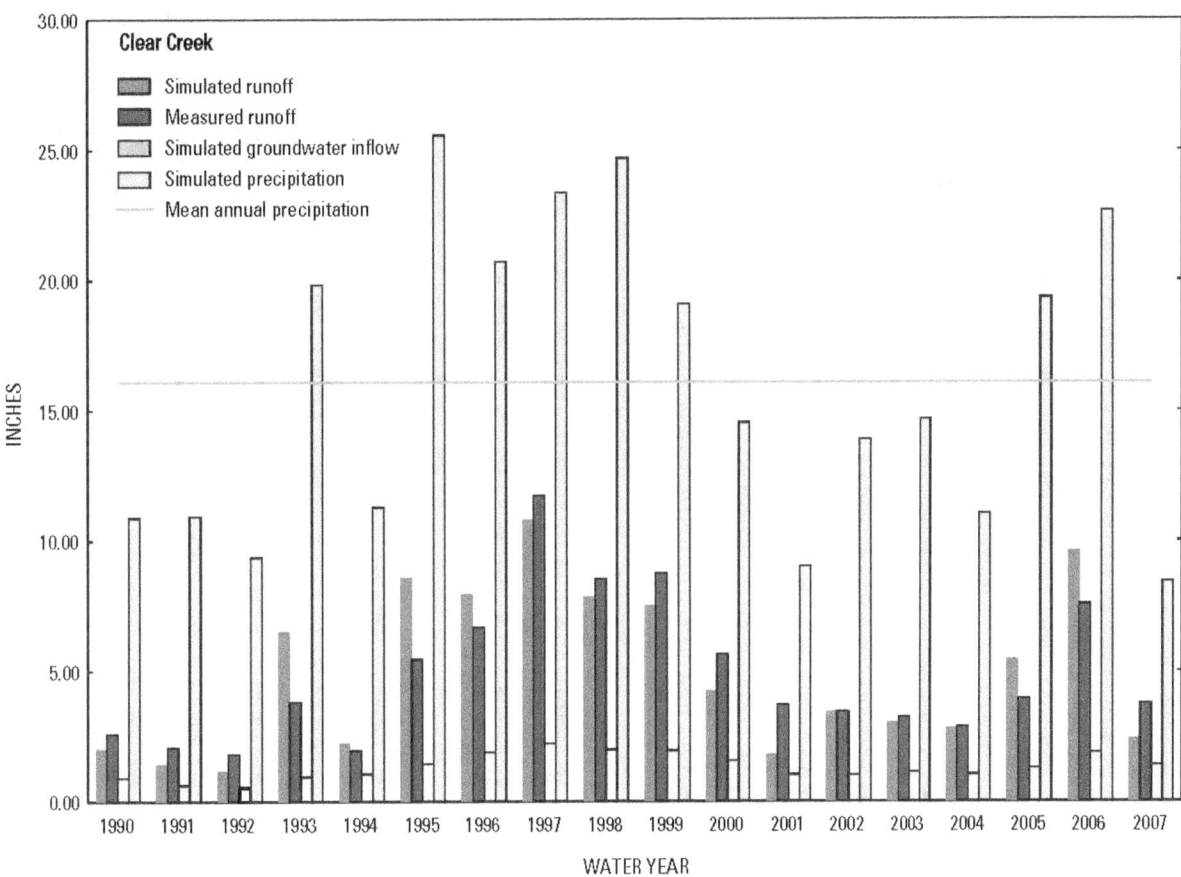

Figure 16.   Simulated and measured annual runoff, simulated groundwater inflow, and precipitation, and mean annual precipitation (blue line) for Clear Creek, Eagle Valley, Nevada, water years 1991–2007.

Comparisons of Water Budget Components to Previous Estimates

Two previous studies by Maurer and others (1996) and Maurer and Berger (1997) provide comparative estimates of groundwater recharge to the Ash Creek and Clear Creek watersheds. Using two techniques, Darcy's Law and the chloride-balance method, estimates of subsurface flow were determined along a hydrogeologic section across the mouth of each canyon near where the streamflow was gaged. The distributions of saturated sediments and fractured bedrock were estimated from test hole and geophysical data along with estimates of hydraulic conductivity and hydraulic gradient. The dissolved chloride concentrations of precipitation, groundwater, and surface water were used to provide an independent estimate of subsurface flow using the chloride-balance method. Precipitation volume was determined from a previously published isohyetal map of precipitation from the period prior to 1978.

From the previous studies, groundwater recharge for Ash Creek was estimated at 200 to 500 acre-ft and for Clear Creek, 1,200 acre-ft. For the present study, mean annual groundwater inflows estimated for the 1980 to 2007 period are similar; Ash Creek is estimated at 490 acre-ft and Clear Creek is estimated at 1,160 acre-ft. This represents about a 7-percent recharge efficiency for Ash Creek and 8 percent for Clear Creek. Estimates of precipitation vary significantly between the earlier and current study, making comparisons of recharge efficiency largely irrelevant. For the present study, mean annual precipitation was estimated at about 7,400 acre-ft for Ash Creek and 14,000 acre-ft for Clear Creek, versus a mean annual precipitation of 8,300 and 23,000 acre-ft, respectively, from the earlier studies. Estimated water yield (the sum of groundwater inflow and surface runoff) volumes are similar for the current and previously described studies, with Ash Creek estimated at about 3,400 acre-ft (compared to the previously estimated 2,800–3,100 acre-ft) and 5,500 acre-ft for Clear Creek for the present study (compared to the previously estimated 5,200 acre-ft). However, due to differences in precipitation volumes between the two studies, the present study estimated the mean annual water yield efficiency (water yield/precipitation) to range from 40 to 46 percent, while the earlier studies estimated Ash Creek between 34 and 37 percent and Clear Creek at around 23 percent of mean annual precipitation.

Ephemeral Watershed Model Results

The middle Carson River ephemeral watersheds selected for simulation included those tributaries to the Carson River downstream from Carson City in the Dayton Valley and Churchill Valley hydrographic areas (fig. 2, table 1). These watersheds are known to have some intermittent runoff based on indirect streamflow measurements (Brunswick Canyon and Sixmile Canyon) or from anecdotal runoff information as provided by field observations by residents and local transportation authorities (written communications, 2009). There is considerably more uncertainty associated with the ephemeral watershed models than with the perennial watershed models due to the lack of continuously gaged steamflow data and the sparse distribution of climate stations. The volumetric storage gage climate data, while not used directly for watershed modeling, indicates more climatic variability than can be captured using the three continuously recording climate stations: Carson City, Virginia City, and Lahontan Dam. Indirect evidence of intermittent flow includes channel incision, field evidence of recent flood debris, and for Churchill and Eldorado Canyons, the presence of riparian zones intermittent to dry sandy reaches. Figure 17 illustrates gaining and loosing stream reaches in Churchill Canyon watershed, as inferred by the presence of intermittent riparian areas (shown in fig. 17 as vegetated areas within Churchill Canyon and Adrian Valley), common in semi-arid environments that typically only flow during spring melt runoff in wetter than normal years, and possibly intermittently during localized summer convective storms.

The overall objective in simulating runoff for the ephemeral watersheds was to simulate some flow during the spring runoff and little to no flow during the summer and fall periods, while maintaining reasonable evapotranspiration volumes (annual volumes ranging from 70 to 90 percent precipitation as inferred from earlier studies (Jeton and Maurer, 2007) and groundwater inflow estimates (recharge efficiency) from 5 to 8 percent of precipitation. Figure 18 is an example of simulated runoff for two adjacent ephemeral watersheds (Brunswick Canyon and Eldorado Canyon) for water years 1992–1997, representing wet and dry years. While the duration of little to no baseflow varies interannually and between the various ephemeral watersheds, as well as the magnitude and timing of runoff peaks, overall the hydrographs for the ephemeral watersheds are similar in seasonal runoff distribution. The ephemeral watersheds exhibit similar vegetation type and canopy density, soils and bedrock geology, and are influenced by the same weather patterns, particularly during the winter when larger scale frontal storms that produce most of the annual precipitation dominate the region. In contrast, the more localized, summer convective patterns affect not only individual watersheds; they often have limited impact within those watersheds, resulting in considerable variability in summer runoff.

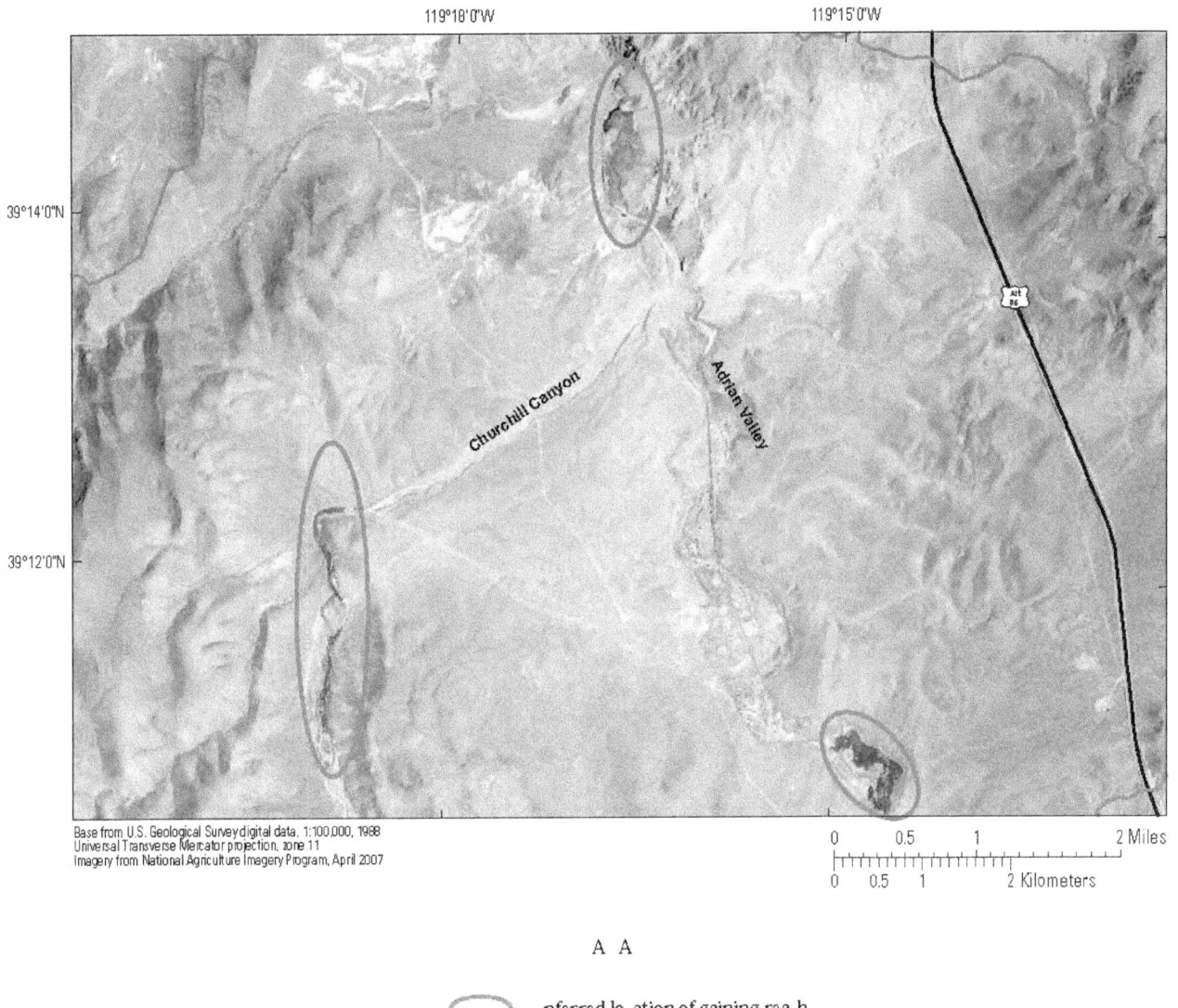

Figure 17. Confluence of Adrian Valley and Churchill Canyon using as the base map a National Agriculture Imagery Program high-altitude aerial photograph, acquired in April 2007, of Churchill Canyon, Nevada.

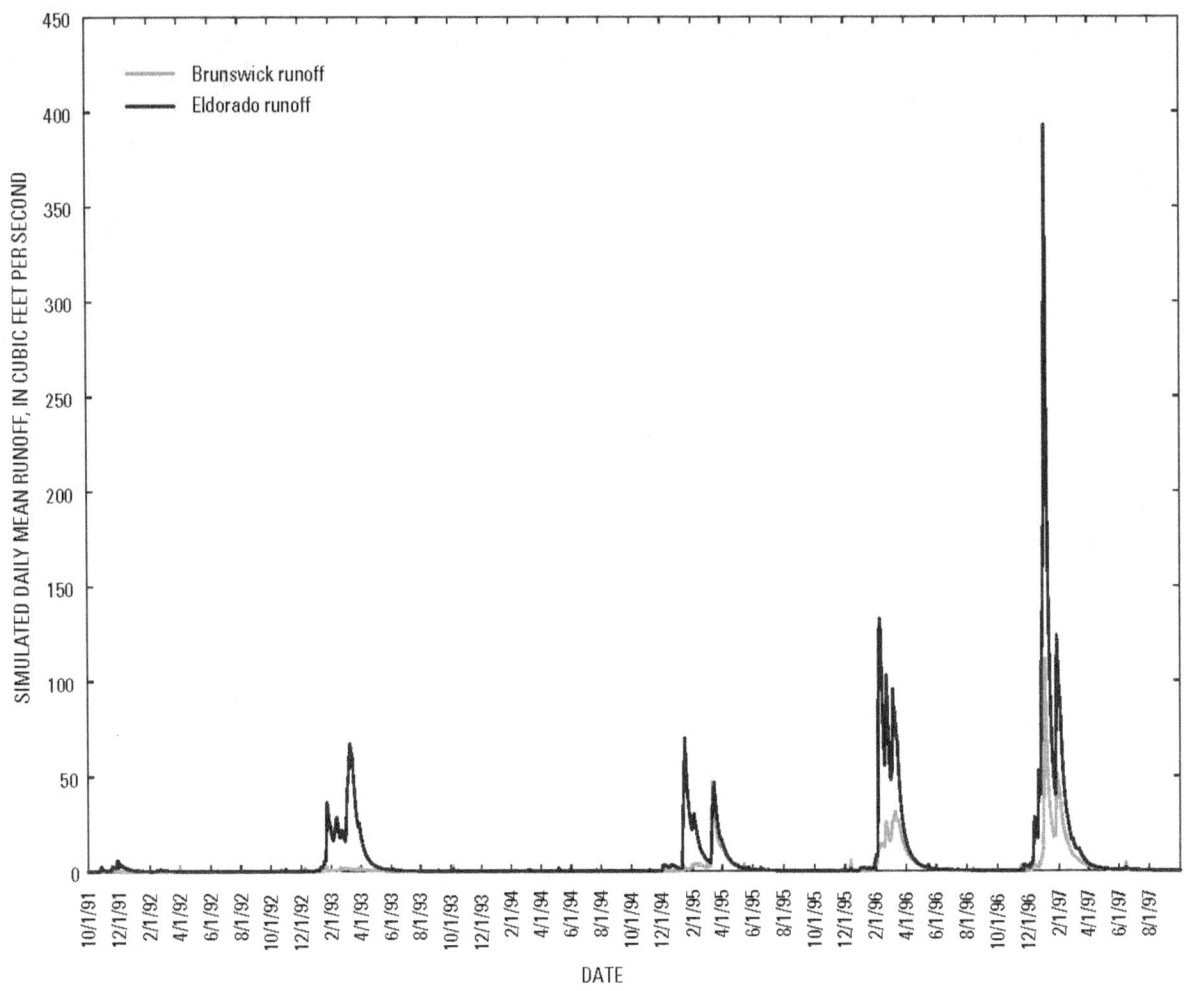

Figure 18. Simulated daily mean runoff from Brunswick and Eldorado Canyon watersheds, Nevada, water years 1992–1997. (Brunswick and Eldorado Canyon watersheds are shown as 3u and 5u, respectively, in figure 2.)

Annual mean water budget histograms provided in this section illustrate the annual variability typical of the climate record used in the simulations. Annual streamflow for the ephemeral watersheds typically follow a pattern of little to no simulated runoff or groundwater inflow for low precipitation years, generally when annual total precipitation is less than 8 in/yr. Immediately following wet periods, simulation results indicate the presence of some carryover subsurface storage into the following year. Photographs of the ephemeral watersheds and their respective streamflow channels are presented and discussed in the following paragraphs to illustrate the semi-arid characteristics of these watersheds. Photograph locations are indexed on figure 19. For discussion purposes, the ephemeral model results are grouped according to both geographic proximity and selection of climate index station(s).

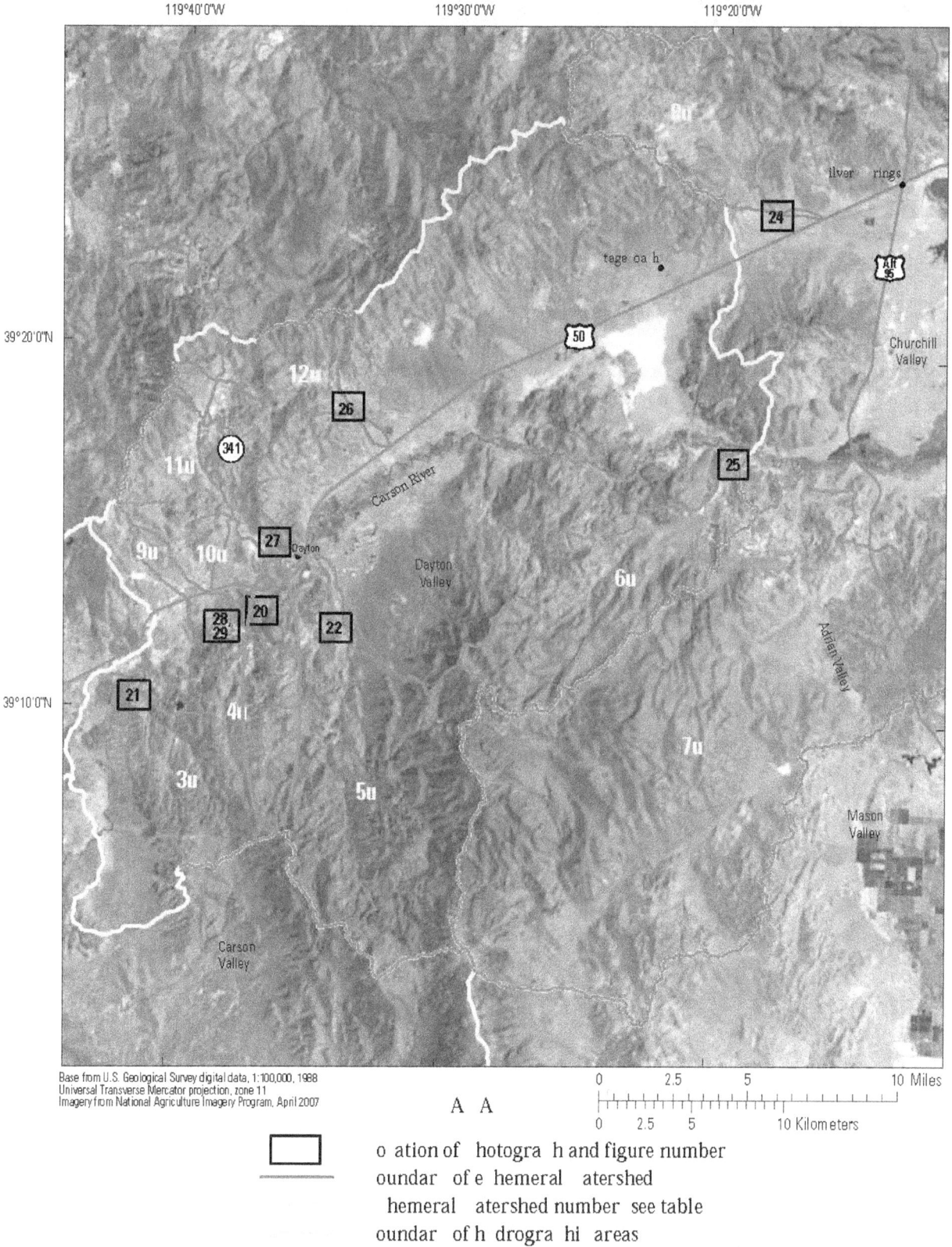

Base from U.S. Geological Survey digital data, 1:100,000, 1988
Universal Transverse Mercator projection, zone 11
Imagery from National Agriculture Imagery Program, April 2007

Figure 19.   Locations of photographs listed in subsequent figures in this report, and ephemeral watersheds, middle Carson River basin, Nevada.

Brunswick, Hackett, and Eldorado Canyon Watersheds

Brunswick Canyon (12.9 mi[2]; fig. 2, 3u) drains into the Carson River downstream from Carson City, bordering both Hackett Canyon (7.15 mi[2]; fig. 2, 4u) to the east and at the upper altitudes shares part of the topographic divide with Eldorado Canyon, the largest of the three watersheds (56 mi[2]; fig. 2, 5u). These watersheds are sparsely vegetated at the lower altitudes by rangeland shrubs and seasonal grasses (fig. 20) with increased stand densities of primarily pinyon-juniper in the upper zones. Brunswick Canyon reflects a typical channel morphology for the study area; sandy alluvial channel bottoms at the lower reaches (fig. 21) and a more bedrock-based channel in the upper reaches. Volcanic and granitic rocks are the dominant rock types. Five indirect flow measurements were recorded for Brunswick Canyon, including a regional flood event in 1986, though none for two other large flood events in water years 1997 and 2006. Indirect estimates were also made for the February 1986 flood in Eldorado Canyon.

All three models used the Carson City climate station for both precipitation and temperature with adjustments made to the precipitation distributed to the higher altitude HRUs. The Carson City climate station (4,651 ft) is situated immediately in the Carson Range rain shadow with precipitation decreasing further east. Precipitation data from the two storage gages, Brunswick Canyon (6,370 ft) and Brunswick Reservoir (5,100 ft), suggest less precipitation in the 6,000 to 7,000 ft zone than would otherwise result from adjusting the Carson City data to fit a distribution suitable for HRUs in the upper

altitude zones, which extend to over 8,000 ft in the Eldorado Canyon watershed. Mean annual precipitation averaged 11 in. basinwide (versus 10.5 in. at the Carson City gage) for both Brunswick and Eldorado Canyons, with a runoff efficiency from 14 to 15 percent for both watersheds and a recharge efficiency ranging from about 4 to 8 percent (table 4). Hackett Canyon is predominately in the 5,000 to 6,000 ft altitude band with a simulated mean annual precipitation of 8 in., a recharge efficiency of 6 percent, and similar runoff efficiency to the adjacent watersheds. Unit of runoff for the three watersheds ranges from 0.09 to 0.14 acre-ft/acre, and the water budget residual from zero to less than 1 percent, suggesting most of the water is accounted for during simulation.

A January 2009 photo of water flowing in the Eldorado Canyon channel (fig. 22) illustrates mid-winter runoff following a brief winter storm several days prior. Simulated annual water budget components for Eldorado Canyon (fig 23) illustrate patterns typical for the area; minimal runoff and groundwater inflow during years when annual precipitation is less than 10 in., and at around 8 in. or less of annual precipitation, evapotranspiration approaches precipitation. For most other (wetter) years, evapotranspiration consumes on average about 76 percent of total precipitation. Mean annual groundwater inflow for the 1978–2007 period for Eldorado Canyon averaged about 2,700 acre-ft with annual volumes ranging from about 30 acre-ft in 2004 (one of the driest years) to a high of over 9,000 acre-ft in 1983, the wettest year in the modeling period.

Figure 20. Lower Hackett Canyon watershed and the Carson River (lower right), January 2009, middle Carson River basin, Nevada.

Figure 21.   Brunswick Canyon, January 2009, middle Carson River basin, Nevada.

Figure 22.   Eldorado Canyon with runoff following an early winter snowmelt, January 2009, middle Carson River basin, Nevada.

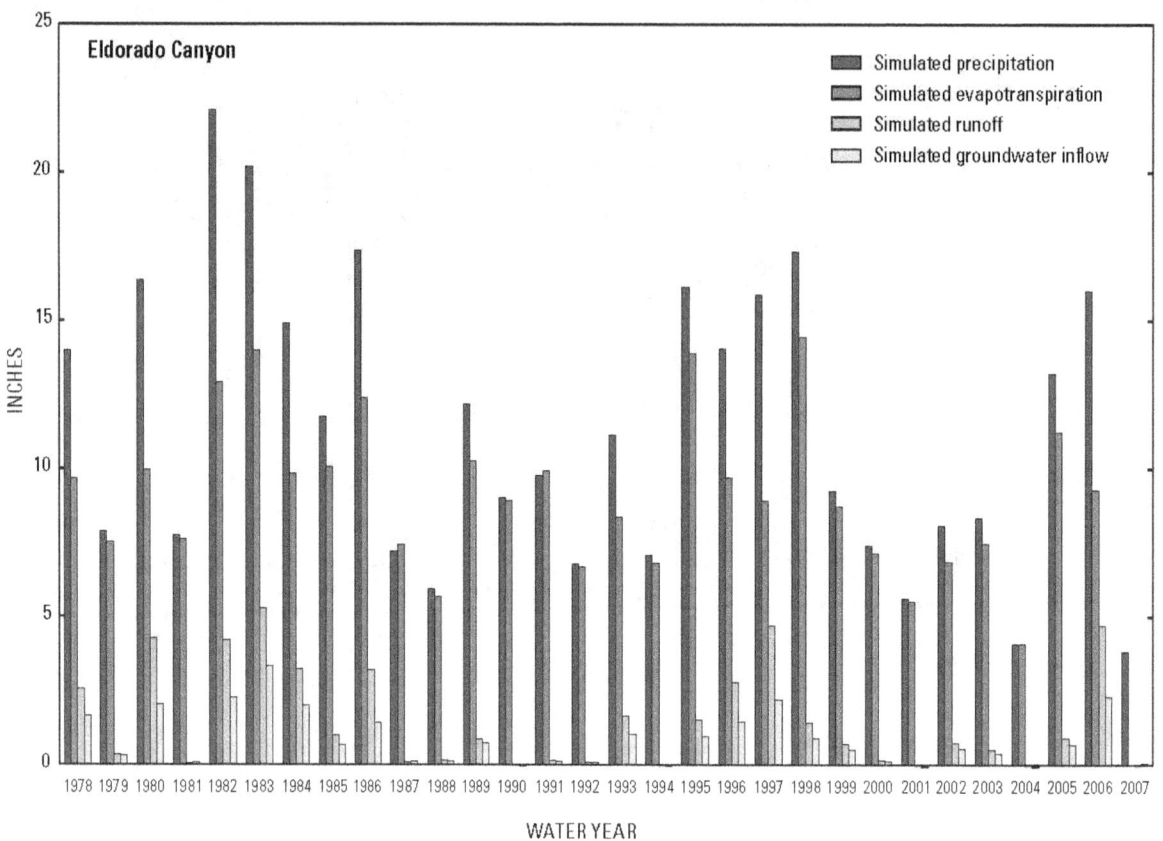

Figure 23.   Simulated annual runoff, precipitation, evapotranspiration, and groundwater inflow for Eldorado Canyon watershed, Nevada, water years 1979–2007.

Churchill, Bull–Mineral, and Ramsey Canyon Watersheds

Churchill Canyon (fig. 2, 7u) is the largest watershed in the study area with an approximately 153-square-mile drainage area. General characteristics include a northeast-southwest trending drainage with the western boundary bordering Eldorado Canyon and a high-altitude catchment area along the eastern crest of the Pine Nut Mountains at altitudes between 7,000 and 9,000 ft. The geology is mainly volcanic and intrusive rock that creates moderate relief within the watershed, and sedimentary and alluvial-fill valleys. As viewed on figure 17, intermittent riparian reaches are along both of the main tributaries, Adrian Valley and Churchill Canyon channels, separated by long reaches of broad, sandy alluvial channels. The eastern highlands of the Pine Nut Mountains are characterized by pinyon-juniper stands interspersed with some rangeland and grassland communities with canopy densities ranging from 50 to over 80 percent. Sparsely populated shrublands cover the valley floor and lowlands with less than 10 percent canopy density.

Bull–Mineral Canyon (29 mi²; fig. 2, 6u) borders Churchill Canyon to the north and drains a narrowly defined, roughly east-west oriented watershed of sparse vegetation and volcanic rock outcropping as barren, talus-laden hillslopes. Bull and Mineral Canyons constitute the two main channels in the watershed. The western upper boundary borders the Pine Nut Mountains crest line between 7,000 and 8,000 ft, with pinyon-juniper stands interspersed with some grassland and shrub. Over half of the watershed area lies at altitudes less than 6,000 ft, covered predominantly with shrubland with less than 10 percent canopy cover. Bull–Mineral and Churchill Canyons were assumed to have similar precipitation and temperature distributions. Channel morphology for Mineral Canyon is complex and may represent a paleohistory of channel migration in response to changing Pleistocene lake levels, and deltaic sedimentary material deposited by a meandering Carson River (Morrison, 1964). While no runoff measurements exist, field evidence of flood debris (not shown) suggest runoff during the region's major flood events, most notably the 1997 and possibly the 2005 flood events. The channel is mainly alluvium with coarse gravel and boulders suggestive of high-energy flow during regional flooding.

In contrast, Ramsey Canyon (24 mi^2; fig. 2, 8u) to the north of Churchill Canyon is more arid with canopy densities less than 15 percent across much of the watershed with the exception of the upper reaches, situated in the 6,000 to 7,000 ft range. Most of the land cover in the watershed is classified as shrubland with localized areas of grassland, a sandy, alluvial channel (fig. 24), exposed volcanic bedrock in the upper reaches within the Flowery Range, and sedimentary units in the zones below 5,000 ft.

All three models used the Lahontan Dam climate station situated at the northeast extent of the study area (fig. 2) with minor adjustments to account for HRU altitude. For the modeling period (water years 1978–2007), simulated mean annual precipitation averaged 9.2 in. for Churchill, 9.4 in. for Bull–Mineral, and 6.2 in. for Ramsey Canyon, situated the furthest east near Lahontan Reservoir and the driest of the ephemeral watersheds. Figure 25 illustrates the annual water budget distribution for Churchill Canyon. Mean annual groundwater inflow for the 1978–2007 period averaged about 5,000 acre-ft with annual volumes ranging from a low of about 30 acre-ft in 2001 (one of the driest years) to a high of over 15,000 acre-ft in 1983 when annual precipitation exceeded 19 in.

Simulated evapotranspiration averaged roughly 85 percent of total precipitation for the three watersheds, leaving little for runoff (between 3 and 11 percent) and between 5 and 7 percent for groundwater inflow (table 4). Ramsey Canyon had no residual water while the Churchill Canyon model residual represented less than 1 percent. In contrast, the higher percentage of residual water for Bull–Mineral Canyon (just over 6 percent) may be attributed to deeper soils and greater subsurface storage. In simulating ephemeral runoff (with emphasis on little to no streamflow during most months of the year) more precipitation and snowmelt is routed to the subsurface reservoir as storage than might be realistic to maintain reasonable precipitation, evapotranspiration, and groundwater inflow rates. When comparing the PRMS mean annual runoff estimates to the Moore runoff estimates (table 4), Ramsey Canyon matched the Moore estimate, while about 40 percent more runoff was simulated using the PRMS model for Churchill Canyon, and 55 percent less runoff was simulated for Bull–Mineral Canyon.

Figure 24. Ramsey Canyon, looking upstream, January 2009, middle Carson River basin, Nevada.

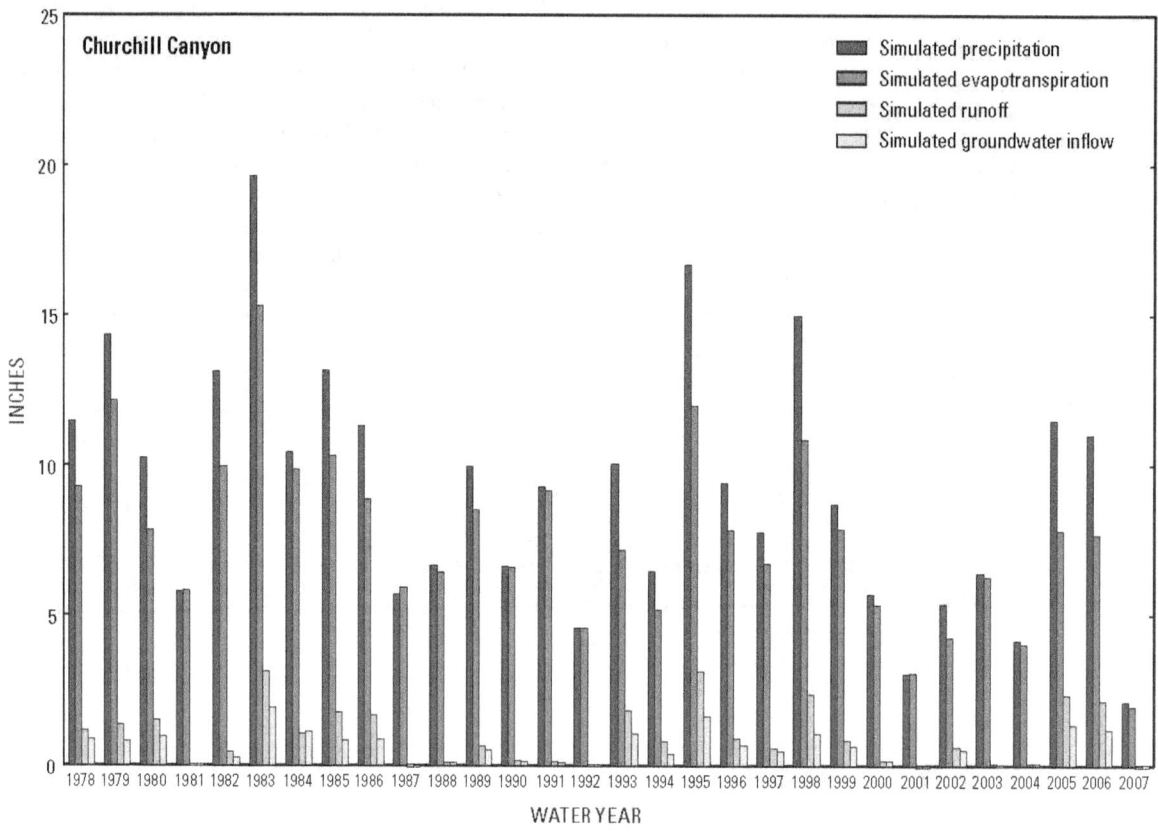

Figure 25.   Simulated annual runoff, precipitation, evapotranspiration, and groundwater inflow for Churchill Canyon watershed, Nevada, water years 1978–2007.

Sixmile, Gold, Daney, and Eureka Watersheds

Four watersheds on the north side of the middle Carson River basin were selected to be modeled based on infrequent yet detectable runoff history. Collectively, these four ephemeral tributaries situated on the eastern slopes of the Virginia and Flowery Ranges have a combined drainage area of approximately 39 mi². Only two indirect measurements exist for Sixmile Canyon, and one for Gold Canyon, which included estimates for the 1986 flood. However, the four watersheds exhibit some evidence of recent flow such as incised channels and flood debris (fig. 26); the latter is most probably from either of the last two major regional floods in 1997 and 2006. Summer convective storms are common in the region and often produce short-duration runoff that may account for more of the channel scour than the less frequent regional winter storms. The four watersheds have similar land-use distribution; shrub and grassland communities, gravelly channel bottoms incised into sparsely vegetated hills, as illustrated in figure 27, and more residential and commercial development than in the previously described watersheds. Urban development, predominantly in the alluvial lowlands, ranges from low density with open space to higher density residential and commercial zones. Development across

the alluvium has altered the natural channel in most of the watersheds, particularly upstream of the canyon mouths. The topographic relief across the alluvial fans is slight and coupled to the urban development in these flatter areas; the channels are less pronounced than up canyon. However, as with Eureka Canyon (figs. 28 and 29), channelization into the volcanic sediments suggests periodic flow most probably during the major flood events already discussed.

Considerable growth has occurred in this part of the Dayton Valley hydrographic area since the late 1960s when Moore developed his methodology based on generalized runoff-altitude relations rather than on land use. Impervious surfaces created as a result of urbanization change not only the character of the hydrograph but also result in higher runoff. The increased impervious area in these watersheds may account for higher mean annual runoff simulated with PRMS, with the exception of Gold Canyon, than runoff estimated using the Moore method; however, no comparable impervious data are available from which to compare with the current period. It may also be that the runoff-altitude relations developed by Moore (1968) may not be suitable for the Virginia Range. Recharge efficiency (table 4) ranged from 3 to 5 percent and runoff efficiency from 7 to 13 percent,

Figure 26. Sixmile Canyon and recent flood debris, looking downstream, January 2009, middle Carson River basin, Nevada.

Figure 27. Gold Canyon watershed, view from ridge overlooking Gold Canyon channel, January 2009, middle Carson River basin, Nevada.

Figure 28.   Eureka Canyon, looking upstream, January 2009, middle Carson River basin, Nevada.

Figure 29.   Eureka Canyon, outlet to the Carson River, January 2009, middle Carson River basin, Nevada.

with Sixmile Canyon simulating the highest mean annual runoff. Mean annual precipitation for these watersheds is low, ranging between 8 and 10 in/yr, leaving little excess water after evapotranspiration. While no measured runoff exists for these watersheds, model parameters were adjusted to simulate ephemeral runoff primarily during the spring melt period for those years with above-average annual precipitation.

Mean annual simulated precipitation for the four watersheds is about 8 in. with a mean evapotranspiration rate of 7 in., indicating very little excess moisture is available for surface runoff or groundwater inflow. Residuals are equal or less than 0.1 percent, indicating most water is accounted for in the model with little to no subsurface storage. Mean annual precipitation at McClellan Peak (7,200 ft) situated near the Gold Canyon crest (fig. 2) is about 9 in. for water years 1997–2007; however, the adjusted-PZM data estimated precipitation for the 7,000-ft range to be about 19 in. The HRU-corrected precipitation used in the watershed models better reflects the mean annual precipitation recorded at McClellan Peak when comparing precipitation totals for the higher altitude HRUs.

Annual water budget components for the four watersheds are comparable (shown only for Sixmile Canyon; fig. 30) with similar interannual variability. General characteristics include little to no runoff or groundwater inflow simulated during years when precipitation is generally below 6 in/yr, when evapotranspiration loss approaches precipitation, and during consecutive wet years when subsurface carryover is cumulative (for example, 1982–83; 1995–98). The 1995–98 period indicates decreasing evapotranspiration with less change in overall precipitation, suggesting cooler air temperatures or earlier-than-usual spring runoff. Mean annual groundwater inflow for the 1978–2007 period for Sixmile Canyon averaged about 400 acre/ft with annual volumes ranging from less than 10 acre/ft in 1992, with annual precipitation less than 5 in. following two previously dry years, to a high of more than 1,400 acre/ft in 1983. For years with above 8 in. of annual precipitation, mean annual ET was roughly 76 percent of precipitation. Overall, recharge and runoff efficiencies for Sixmile Canyon were 5 and 13 percent, respectively.

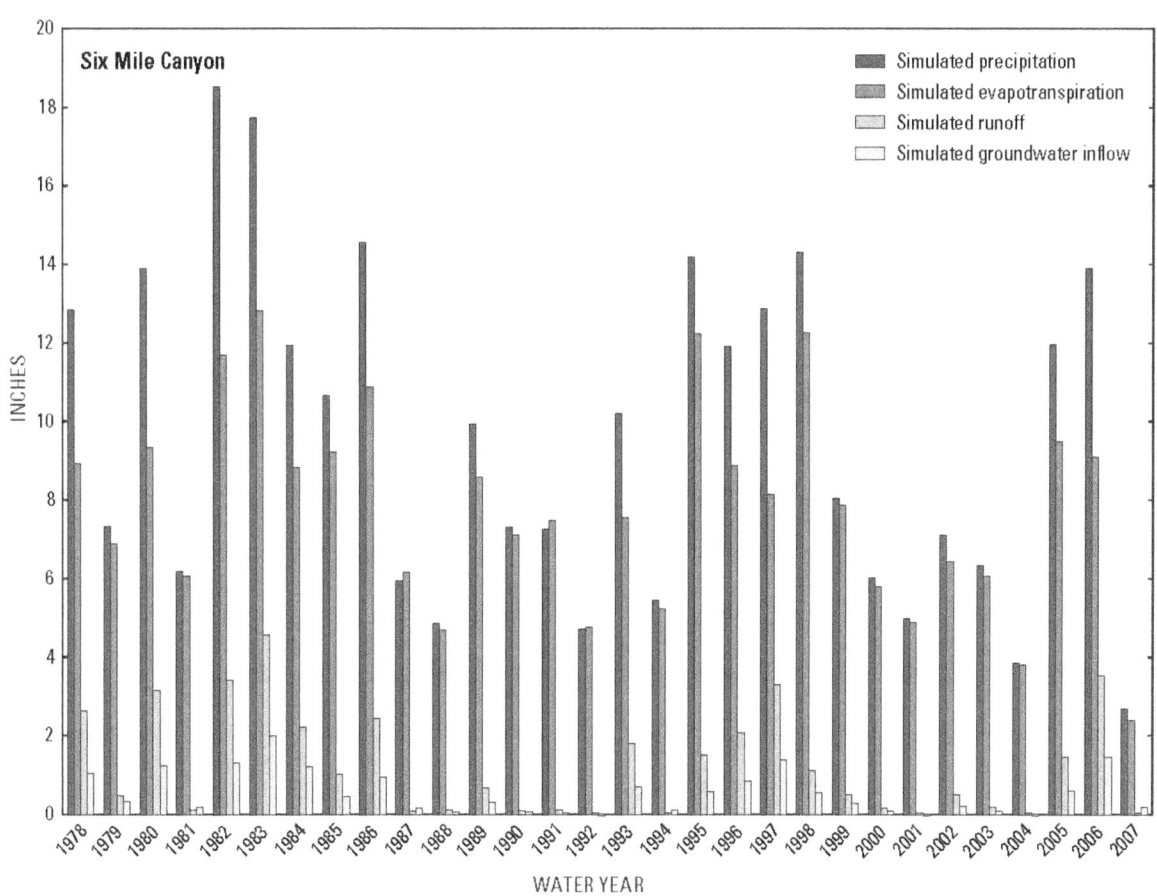

Figure 30.   Simulated annual runoff, precipitation, evapotranspiration, and groundwater inflow for Sixmile Canyon watershed, Nevada, water years 1978–2007.

Model and Data Limitations

The precipitation-runoff model is a mathematical representation of the physical processes that occur in the watershed. The quality of the model results depend on the accuracy of the representation of the physical processes (model error), the quality and accuracy of the precipitation and air-temperature input time series and runoff calibration time series (data error), and the accuracy of the calibrated model parameters relative to physical watershed processes (parameter error; van Heeswijk, 2006).

Those error sources most affecting the watershed models for the middle Carson River basin include: the lack of streamflow data for much of the study area, the assumption that the ephemeral watersheds are hydrologically similar to one another (particularly affecting groundwater inflow rates), the adequacy of available climate data and the accuracy of precipitation estimates using an adjusted PZM distribution (Maurer and others, 2009), the coarseness of the PET data used to set the calibration, the scale and suitability of soil and vegetation density data, and the sensitivity of the model in simulating baseflow when using the PRMS groundwater sink parameter to estimate groundwater inflow (particularly during years of below normal precipitation). Knowing recharge rates for the different hydrogeologic units (and by extension the HRUs) within each modeled watershed might improved the spatial distribution of groundwater inflow (rather than relying on the PRMS groundwater sink term used to represent a watershed-wide summation of potential groundwater inflow at the outlet of the watershed).

Watersheds are dynamic systems. Land cover type, canopy cover density, and the percentage of impervious area are static parameters in PRMS. For this study, land cover conditions reflect characteristics for 2001, when the digital maps were compiled, while the period of record used for modeling spans water years 1978–2007. Canopy cover densities affect primarily the computation of the longwave and shortwave radiation components of the snow energy balance, thereby affecting snowpack melt rates. Specifically, canopy cover densities affect the longwave radiation exchange between vegetation canopy and the snowpack surface, and the net shortwave radiation term for transmission coefficient for the winter cover density over the snowpack. A limited amount of water is lost through canopy interception, determined by the plant canopy density for both summer and winter cover across the HRU area. Urban land-use designation affects the acreage assumed to be impervious, thereby affecting surface runoff and soil infiltration.

The scale of the soil data limits the extent to which the watershed models can represent the actual hydrologic system. The STATSGO soils data are mapped at a scale of 1:250,000, resulting in a 1,000-meter grid resolution and a minimum mapping unit of 1,544 acres. Soil attributes used in simulating soil-zone infiltration and deeper percolation are coarse at this resolution and generally reflect broad properties of the parent material. The soil parameters influence the distribution of water between the surface and subsurface reservoirs and ultimately affect the distribution of interflow, baseflow, and groundwater inflow. Soil-water holding capacity was adjusted upward for most HRUs to better simulate ephemeral runoff while estimating reasonable groundwater inflow. A generalized permeability rating from a 1:500,000 hydrogeology map (Stewart and Carlson, 1978) was used in a qualitative manner in concert with the STATSGO-derived soil characteristics to assist in adjusting soil storage parameters. In addition, the amount of actual evapotranspiration is influenced by the generalized PRMS soil designation of sand, loam, or clay (derived from STATSGO) and the ratio of available water to the maximum soil-water storage at a given simulation step.

Simulated streamflow is a composite of surface runoff, shallow subsurface flow, and groundwater flow. For the perennial watersheds with continuously gaged streamflow data, the dominance of one streamflow component over the other influences the shape of the simulated hydrograph. The simulated groundwater inflow is set at a constant rate in the model for the selected modeling period. In low runoff years, less water is routed to the subsurface reservoirs and thus less water is available for baseflow and groundwater inflow. When running PRMS as a stand-alone model, not dynamically linked to a groundwater model, determining groundwater inflow by simulating the removal of water from the subsurface reservoirs can result in a tendency to underestimate baseflow for the drier years, particularly when adjusting the model to fit snowmelt runoff. Propagated for several consecutively low precipitation years (for example, from 2000 to 2004), the tendency to underestimate baseflow can potentially increase the modeling error for the period of record.

The rain shadow effect of the Carson Range influences precipitation in the middle Carson River basin as much as altitude. The Marlette Lake SNOTEL (fig. 2) adequately represents precipitation at the crest line while the stations to the east, from the Carson City station to the Lake Lahontan station, are increasingly influenced by the Carson Range rain shadow effect. When comparing precipitation amounts between the Carson City gage and the Dead Camel and Lahontan sites (all three in the mid 4,000 ft range; fig. 7), the rain shadow effect is evident. For the current study, the adjusted-PZM overestimated precipitation for HRUs above 6,500 ft in the Dayton and Churchill Valley hydrologic areas when compared to station data. When comparing the adjusted-PZM data in the Sixmile Canyon watershed to mean annual precipitation at the Virginia City climate site, HRUs at altitudes above 6,000 ft had up to 60 percent more mean annual precipitation when using the PZM estimates. For this reason, precipitation correction factors for HRUs above 6,000 ft were adjusted to better reflect the Virginia City station data rather than the adjusted-PZM ratio. The neighboring storage gage on McClellan Peak recorded even less mean annual precipitation than the Virginia City gage, again suggesting the adjusted-PZM data overestimated precipitation at least in the Virginia and Flowery Ranges. To illustrate, the observed daily precipitation at the index station was modified using the adjusted-PZM ratio for each HRU, and then adjusting (decreasing) the runoff and (increasing) the subsurface flow parameters to simulate ephemeral runoff. The result was an increase in groundwater inflow to well above 10 percent of total precipitation, higher than the previously estimated runoff efficiencies, suggesting excess precipitation input to the model using the adjusted-PZM ratio.

Daily air temperature was used in the computation of evaporation, transpiration start and end dates, sublimation, precipitation form, and snowmelt. Point or station data such as the precipitation data are extrapolated to each HRU using a set of monthly adjustment coefficients. The selection of a climate station was based primarily on geographic proximity. Like precipitation, in the absence of a high- and low-altitude set of temperature stations, one of the four index stations was assigned to each modeled area (table 1). Unlike precipitation distribution, air temperature data exhibits little interannual and regional variation, lending less uncertainty in the daily HRU temperature estimates than for precipitation. However, error may be introduced when using one temperature station and static monthly temperature lapse rates to adjust for differences in altitude.

The streamflow time series used to calibrate the two perennial watersheds, Ash Canyon and Clear Creek, have some uncertainty associated with the data, less related to direct measurement error then to upstream diversions or withdrawals, or out-of-basin groundwater flow. The few indirect measurements available for some of the ephemeral watersheds can be relied upon less as data used to match individual modeled events then as general evidence that these watersheds most probably yield some runoff during region-wide snowstorms. Most of the indirect measurements are rated as "fair" to "poor," a reflection of both inherent measurement error and the quality of geomorphic flood evidence. Lacking measured streamflow data, calibration of the ephemeral watershed models was limited to adjustment of precipitation and evapotranspiration, and reasonable estimates of groundwater inflow obtained from earlier studies. Moore's estimates provided a coarse comparison of runoff volume; however, when applying these regional estimates to watershed areas such as those modeled in the present study where local geology, precipitation, vegetation, and land use may vary, the regional runoff-altitude relations may no longer be suitable for direct comparison.

The lack of streamflow data, particularly downstream of Eagle Valley, and the ephemeral nature of runoff in the tributaries selected for simulation resulted in the manipulation of particularly the subsurface flow and soil parameters within the PRMS model. The coarseness of the STATSGO data allowed for greater adjustment of these parameters to successfully simulate a plausible ephemeral hydrograph, whereby surface runoff is concentrated during the spring snowmelt period with little to no baseflow prior to or after this period. Once the computed precipitation and ET volumes were reasonable, the focus was to limit runoff to the snowmelt period while maintaining what was considered to be (from

prior studies) reasonable groundwater inflow. The PRMS model is a stand-alone computer program that does not simulate groundwater inflow in a physical sense but rather allows the user to allocate a portion of the subsurface storage to deep percolation. While simulating the proportion of infiltrated water routed as groundwater inflow, the immediate effect is to decrease the water available for the baseflow component of total streamflow. This is particularly evident during below normal precipitation years when the baseflow contribution to streamflow is more pronounced due to less snowmelt runoff. However, for the present study, simulating reasonable groundwater inflow volumes was a higher priority than matching the baseflow component of the hydrograph.

Summary and Conclusions

To address concerns about the uncertainties that future land- and water-use practices may have on flow in the Carson River, the U.S. Geological Survey, in cooperation with the Bureau of Reclamation, began a study in 2008 to develop a numerical model to simulate groundwater and surface-water interactions in the Carson River upstream from Lahontan Dam and downstream from Carson Valley. As part of this study, watershed models were developed to provide estimates of runoff tributary to the Carson River and the potential for groundwater inflow. The Precipitation-Runoff Model System (PRMS) model is a physically based, distributed-parameter model where the spatial variability of land characteristics that affect runoff and groundwater inflow (recharge) are accounted for by discretizing the watershed into hydrologic response units (HRUs). A geographic information system, the Weasel toolbox, was used to manage spatial data, to digitally characterize model drainages, and to develop 300-meter grid cell HRUs. PRMS models were developed for 2 perennial watersheds in Eagle Valley—Ash Canyon Creek and Clear Creek—and 10 ephemeral watersheds in the Dayton and Churchill Valley hydrologic areas—Brunswick Canyon, Bull–Mineral Canyon, Churchill Canyon, Daney Canyon, Eldorado Canyon, Eureka Canyon, Gold Canyon, Hackett Canyon, Ramsey Canyon, and Sixmile Canyon. Initial parameter values for the modeled watersheds were derived from a previously published study for perennial and ephemeral drainages in the Carson Valley area.

Model calibration was constrained by daily mean flows for the two perennial watersheds in Eagle Valley, and mean annual runoff estimates for the ephemeral drainages. The calibration periods were water years 1980–2007 for Ash Canyon Creek, and 1991–2007 for Clear Creek, allowing for one year prior to the calibration period for parameter initialization. Simulations for the full modeling record from water years 1978 to 2007 were then made for the 10 ephemeral watersheds and from 1980 to 2007 for both perennial watersheds, and their respective mean annual water budget components were computed.

The watershed models were affected by (1) the assumption that the ephemeral watersheds are basically hydrologically similar to one another, (2) the scale of the soil and land cover data used, (3) the adequacy of available climate data and the accuracy of precipitation estimates using modified linear-relations, and (4) the suitability of mean annual runoff estimates derived from earlier studies. Daily data from four continuously recording climate stations were used as input to the watershed models. For the two perennial watershed models, precipitation adjustments reflected orographic differences in HRU and climate station altitude for low- and high-altitude stations. For the ephemeral areas, daily precipitation was adjusted initially on an existing gridded, mean annual precipitation data set derived from regressional analyses for western Nevada. Comparisons to precipitation totals from several storage precipitation gages in the study area found the gridded data set to overestimate precipitation in the higher altitude regions of most of the ephemeral modeled watersheds, possibly due to underestimation of the rain shadow effect from the Carson Range. While uncertainty is present when modeling ephemeral watersheds, regional evapotranspiration and groundwater inflow rates, and some indirect runoff estimates, guided the amount of precipitation and subsequent runoff and groundwater flow considered to be reasonable. Mean annual runoff and recharge efficiencies were computed for each modeled watershed to evaluate the distribution of precipitation within the surface runoff and subsurface flow components of the annual water budget.

Overall bias for Ash Canyon Creek was satisfactory for the seasonal aggregates and only slightly higher at about 6 percent for the mean monthly and mean annual runoff. Modeling error was lowest for the fall season and highest for the summer season, with a mean annual runoff efficiency of around 40 percent. However, when viewed in the context of mean monthly flow, computed as a percentage of annual flow, the monthly distribution was reasonable for all months. A mean annual recharge efficiency of 7 percent was only slightly higher than previous estimates. Modeling results for the Clear Creek model indicate a high bias towards oversimulating the summer and undersimulating the fall seasonal aggregates, with reasonable bias and error for the winter and spring seasons. Agricultural diversions upstream from the gage may account for some of the summer discrepancy, while for the fall season the cessation of evapotranspiration resulting in return flow was not captured by the model, particularly in simulating evapotranspiration from riparian communities that were not represented in the model. There may also be out-of-basin flow, as suggested in earlier studies, that would affect the baseflow component of runoff. Mean annual groundwater inflow for the 1978–2007 period averaged about 480 acre-feet for the Ash Canyon Creek and 1,160 acre-feet for Clear Creek watersheds, with considerable interannual variation. Model results for groundwater inflow were consistent with previous studies; however, estimates of mean annual precipitation vary significantly. More than twice the annual precipitation

was estimated for the Clear Creek watershed in the previous study, precluding direct comparisons of recharge and runoff efficiencies and water-yield estimates derived in the present study.

For the ephemeral watersheds, estimates of recharge efficiency ranged from 3 to 8 percent, and for runoff efficiency, from 3 to 15 percent. The volume of mean annual groundwater inflow ranged from about 40 acre-feet for Eureka Canyon to just less than 5,000 acre-feet for Churchill Canyon for the 1978–2007 period of simulation. The volume of annual groundwater inflow differs considerably within each watershed, although similar patterns between watersheds reflect modeling periods with marked climatic variability. In addition, annual groundwater inflow volumes, as simulated, are affected by some carryover of subsurface storage from previous years with higher precipitation, as evidenced in the 1982–86 and 1993–98 water years.

Estimates of mean annual runoff were independently computed for the 10 ephemeral watersheds using regional altitude-runoff relations developed by Moore (1968) for Nevada. While these estimates were considered in a general manner when evaluating the current water budget components, no adjustments were made to the models given the coarse nature of this data, and the changes in land use over time that are presumed to have affected both surface runoff and groundwater inflow. Nonetheless, comparison of the Moore runoff estimates to PRMS-simulated runoff indicates a wide range of runoff differences, from a relatively close fit for Brunswick and Ramsey Canyons to more than a 50-percent difference for Hackett, Bull–Mineral, Sixmile, Daney, and Eureka Canyons. The latter two watersheds had the largest differences (at 200 percent), which may be partially attributed to increased impervious cover due to land-use changes in recent decades, from largely rangeland to light to mid-density urban development. The Moore method may be more applicable to larger regional areas rather than as applied to individual watersheds of the size modeled in this study.

Given the similarity in altitude distribution and physiographic characteristics of both the perennial and ephemeral watersheds, the hydrologic characteristics as modeled in this study could be transferred to those as yet unmodeled parts of the middle Carson River basin. Ash Canyon Creek and Clear Creek models represent the runoff characteristics of the watersheds draining the east slope of the Carson Range in Eagle Valley, while the 10 ephemeral watersheds represent the general characteristics of the more arid and more sparsely vegetated Dayton Valley and Churchill Valley hydrologic areas. Annual water budgets for both the perennial and ephemeral watersheds indicate significant interannual variability in runoff and groundwater inflow is caused by climate variations. The watersheds respond differently not only to above- and below-average precipitation years but to whether extreme climate patterns (particularly dry years) persist for extended (multiple year) periods, thereby depleting any residual subsurface storage.

References Cited

Anderson, E.A., 1973, National Weather Service River Forecast System—Snow accumulation and ablation model: U.S. Department of Commerce, NOAA Technical Memorandum NWS–Hydro–17, March 1973.

Beven, K.J., 2001, Rainfall-runoff modeling—The primer: New York, John Wiley & Sons, 360 p.

Brown and Caldwell, 2004, Silver Springs groundwater evaluation prepared for the Carson Water Subconservancy District, Carson City, Nevada: Carson City, Nevada, Brown and Caldwell, 13 p.

Cardinalli, J.L., Roach, L.M., Rush, F.E., and Vasey, B.J., 1968, State of Nevada hydrographic areas, in Rush, F.E., Index of hydrographic areas of Nevada: Nevada Division of Water Resources Information Report 6, 33 p.

Farnsworth, R.K., Thompson, E.S., and Peck, E.L., 1982, Evaporation atlas for the contiguous 48 United States: Washington, D.C., U.S. Department of Commerce, NOAA Technical Report NWS 33, 4 pls., 26 p.

Frank, E.C., and Lee, R., 1966, Potential solar beam irradiation on slopes: U.S. Department of Agriculture, Forest Service Research Paper RM–18, 116 p.

Garcia, K.T, Munson, R.H, Spaulding, R.J. and Vasquez, S.L., 2002, Water Resources Data, Nevada, Water Year 2001, U.S. Geological Survey Water-Data Report NV-01-1, p. 528.Haan, C.T., Johnson, H.P., and Brakensiek, D.L., 1982, Hydrologic modeling of small watersheds: American Society of Agricultural Engineers, ASAE Monograph no. 5, 533 p.

Harrill, J.R., and Preissler, A.M., 1994, Ground-water flow and simulated effects of development in Stagecoach Valley, a small, partly drained basin in Lyon and Storey Counties, western Nevada: U.S. Geological Survey Professional Paper 1409-H, 74 p.

Jensen, M.F., and Haise, H.R., 1963, Estimating evapotranspiration from solar radiation: American Society of Agricultural Engineers: Journal of Irrigation and Drainage, v. 89, no. IR4, p. 15041.

Jensen, M.E., Rob, D.C., and Franzoy, C.E., 1969, Scheduling irrigations using climate-crop-soil data: American Society of Civil Engineers, National Conference on Water Resources Engineering, Proceedings, 20 p.

Jeton, A.E., 1999a, Precipitation-runoff simulations for the Lake Tahoe basin, California and Nevada: U.S. Geological Survey Water-Resources Investigations Report 99–4110, 61 p.

Jeton, A.E., 1999b, Precipitation-runoff simulations for the upper part of the Truckee River basin, California and Nevada: U.S. Geological Survey Water-Resources Investigations Report 99–4282, 41 p.

Jeton, A.E., and Maurer, D.K., 2007, Precipitation and runoff simulations of the Carson Range and Pine Nut Mountains, and updated estimates of ground-water inflow and the ground-water budget for basin-fill aquifers of Carson Valley, Douglas County, Nevada, and Alpine County, California, U.S. Geological Survey Scientific Investigations Report 2007–5205, 55 p. (Also available at http://pubs.usgs.gov/sir/2007/5205/.)

Jeton, A.E., Dettinger, M.D., and Smith, J.L., 1996, Potential effects of climate change on streamflow, eastern and western slopes of the Sierra Nevada, California and Nevada: U.S. Geological Survey Water-Resources Investigations Report 95-4260, 44 p.

Joung, H., Trimmer, J.H., and Jewell, R., 1983, Nevada watershed studies, 1963–1980, U.S. Department of the Interior, Bureau of Land Management, Nevada State Office, 447 p.

Koczot, K.M., Jeton, A.E., McGurk, B.J., and Dettinger, M.D., 2005, Precipitation-runoff processes in the Feather River basin, northeastern California, and streamflow predictability, water years 1971–97, U.S. Geological Survey Scientific Investigations Report 2004–5202, 82 p. (Also available at http://pubs.usgs.gov/sir/2004/5202/.)

Leavesley, G.H., Litchy, R.W., Troutman, M.M., and Saindon, L.G., 1983, Precipitation-runoff modeling system—User's manual: U.S. Geological Survey Water-Resources Investigation Report 83–4238, 207 p.

Lopes, T.J., and Medina, R.L., 2007, Precipitation zones of west-central Nevada, Journal of the Nevada Water Resources Association, v. 4, no. 2, 19 p.

Maurer, D.K., 2011, Geologic framework and hydrogeology of the middle Carson River basin, Eagle, Dayton, and Churchill Valleys, west-central Nevada: U.S. Geological Survey Scientific Investigations Report 2011-5055, 62 p. (Also available at http://pubs.usgs.gov/sir/2011/5055/.)

Maurer, D.K., and Berger, D.L., 1997, Subsurface flow and water yield from watersheds tributary to Eagle Valley hydrographic area, west-central Nevada: U.S. Geological Survey Water-Resources Investigations Report 97–4191, 56 p.

Maurer, D.K., and Berger, D.L., 2007, Water budgets and potential effects of land and water-use changes for Carson Valley, Douglas County, Nevada, and Alpine County, California: U.S. Geological Survey Scientific Investigations Report 2006–5305, 61 p. (Also available at http://pubs.usgs.gov/sir/2006/5305/.)

Maurer, D.K., Berger, D.L., Paul, A.P., and Mayers, C.J., 2009, Analysis of streamflow trends, ground-water and surface-water interactions, and water quality in the upper Carson River basin, Nevada and California: U.S. Geological Survey Scientific Investigations Report 2008–5238, 192 p. (Also available at http://pubs.usgs.gov/sir/2008/5238/.)

Maurer, D.K., Berger, D.L., and Prudic, D.E., 1996, Subsurface flow to Eagle Valley from Vicee, Ash, and Kings Canyons, Carson City, Nevada, estimated from Darcy's law and the chloride-balance method: U.S. Geological Survey Water-Resources Investigations Report 96–4088, 74 p.

Maurer, D.K., Lopes, T.J, Medina, R.L., and Smith, J.L. 2004, Hydrogeology and hydrologic landscape regions of Nevada: U.S. Geological Survey Scientific Investigations Report 2004–5131, 35 p. (Also available at http://pubs.usgs.gov/sir/2004/5131/.)

Moore, D.O., 1968, Estimating mean streamflow in ungaged semiarid areas: Nevada Department of Conservation and Natural Resources, Water Resources Bulletin 36, 11 p.

Moore, J.G., 1969, Geology and mineral deposits of Lyon, Douglas, and Ormsby Counties, Nevada: Nevada Bureau of Mines and Geology Bulletin 75, 45p.

Morrison, R.B., 1964, Lake Lahontan—Geology of southern Carson Desert, Nevada: U.S. Geological Survey Professional Paper 401, 156 p.

Obled, C., and Rosse, B.B., 1977, Mathematical models of a melting snowpack at an index plot: Journal of Hydrology, v. 32, no. 1–2, p. 139–163.

Stewart, J.H. and Carlson, J.E., 1978, Geologic map of Nevada: U.S. Geological Survey, prepared in cooperation with the Nevada Bureau of Mines and Geology, 1:500,000 scale, 2 sheets.

Swift, L.W., 1976, Algorithm for solar radiation on mountain slopes: Water Resources Research, v. 12, no. 1, p. 108–112.

U.S. Department of Agriculture, 1991, Natural Resources Conservation Service, National Soil Survey Center, State Soil Geographic (STATSGO) Data Base, Miscellaneous publication Number 492. Available from http://soils.usda.gov/survey/geography/statsgo/.

U.S. Department of Agriculture, 2008, Natural Resources Conservation Service, National Water and Climate Center, accessed from http://www.wcc.nrcs.usda.gov/snow/about.html.

U.S. Geological Survey, 1999, National Elevation Dataset: U.S. Geological Survey data, accessed November 1, 2010, at http://gisdata.usgs.net/ned.

van Heeswijk, Marijke, 2006, Development of a precipitation-runoff model to simulate unregulated streamflow in the Salmon Creek basin, Okanogan County, Washington: U.S. Geological Survey Scientific Investigations Report 2006–5274, p. 29. (Also available at http://pubs.usgs.gov/sir/2006/5274/.)

Western Regional Climate Center, Desert Research Institute, Historical Climate Information, 2008, accessed September 2008, at http://www.wrcc.dri.edu/summary/Climsmnv.html.

Viger, R.J., 2008, An overview of the GIS Weasel: U.S. Geological Survey Fact Sheet 2008–3004, 2 p. (Also available at http://pubs.usgs.gov/fs/2008/3004/.)

Viger, R.J., and Leavesley, G.H., 2007, The GIS Weasel user's manual, U.S. Geological Survey Techniques and Methods book 6, chap. B4, 201 p. (Also available at http://pubs.usgs.gov/tm/2007/06B04/.)